Advances in

Geosciences

Volume 18: Ocean Science (OS)

ADVANCES IN GEOSCIENCES

Editor-in-Chief: Wing-Huen Ip *(National Central University, Taiwan)*

A 5-Volume Set

Volume 1: Solid Earth (SE)
ISBN-10 981-256-985-5
Volume 2: Solar Terrestrial (ST)
ISBN-10 981-256-984-7
Volume 3: Planetary Science (PS)
ISBN-10 981-256-983-9
Volume 4: Hydrological Science (HS)
ISBN-10 981-256-982-0
Volume 5: Oceans and Atmospheres (OA)
ISBN-10 981-256-981-2

A 4-Volume Set

Volume 6: Hydrological Science (HS)
ISBN-13 978-981-270-985-1
ISBN-10 981-270-985-1
Volume 7: Planetary Science (PS)
ISBN-13 978-981-270-986-8
ISBN-10 981-270-986-X
Volume 8: Solar Terrestrial (ST)
ISBN-13 978-981-270-987-5
ISBN-10 981-270-987-8
Volume 9: Solid Earth (SE), Ocean Science (OS)
& Atmospheric Science (AS)
ISBN-13 978-981-270-988-2
ISBN-10 981-270-988-6

A 6-Volume Set

Volume 10: Atmospheric Science (AS)
ISBN-13 978-981-283-611-3
ISBN-10 981-283-611-X
Volume 11: Hydrological Science (HS)
ISBN-13 978-981-283-613-7
ISBN-10 981-283-613-6
Volume 12: Ocean Science (OS)
ISBN-13 978-981-283-615-1
ISBN-10 981-283-615-2
Volume 13: Solid Earth (SE)
ISBN-13 978-981-283-617-5
ISBN-10 981-283-617-9
Volume 14: Solar Terrestrial (ST)
ISBN-13 978-981-283-619-9
ISBN-10 981-283-619-5
Volume 15: Planetary Science (PS)
ISBN-13 978-981-283-621-2
ISBN-10 981-283-621-7

A 6-Volume Set

Volume 16: Atmospheric Science (AS)
ISBN-13 978-981-283-809-4
ISBN-10 981-283-809-0
Volume 17: Hydrological Science (HS)
ISBN-13 978-981-283-811-7
ISBN-10 981-283-811-2
Volume 18: Ocean Science (OS)
ISBN-13 978-981-283-813-1
ISBN-10 981-283-813-9
Volume 19: Planetary Science (PS)
ISBN-13 978-981-283-815-5
ISBN-10 981-283-815-5
Volume 20: Solid Earth (SE)
ISBN-13 978-981-283-817-9
ISBN-10 981-283-817-1
Volume 21: Solar Terrestrial (ST)
ISBN-13 978-981-283-819-3
ISBN-10 981-283-819-8

EDITORS

Editor-in-Chief: Wing-Huen Ip

Volume 16: Atmospheric Science (AS)
Editor-in-Chief: Jai Ho Oh
Editors: G. P. Singh
                    C. C. Wu
                    K.-J. Ha

Volume 17: Hydrological Science (HS)
Editor-in-Chief: Namsik Park
Editors: Ji Chen
                    Joong-Hoon Kim
                    Jinping Liu
                    Young-Il Moon
                    Sanjay Patil
                    Ashok Kumar Rastogi
                    Simon Toze

Volume 18: Ocean Science (OS)
Editor-in-Chief: Jianping Gan
Editors: Minhan Dai
                    Anne Mueller
                    Murty Vadiamani

Volume 19: Planetary Science (PS)
Editor-in-Chief: Anil Bhardwaj
Editors: Yasumasa Kasaba
                    Guillermo Manuel Muñoz Caro
                    Takashi Ito
                    Paul Hartogh
                    C. Y. Robert Wu
                    S. A. Haider

Volume 20: Solid Earth (SE)
Editor-in-Chief: Kenji Satake

Volume 21: Solar & Terrestrial Science (ST)
Editor-in-Chief: Marc Duldig
Editors: P. K. Manoharan
                    Andrew W. Yau
                    Q.-G. Zong

REVIEWERS

The Editors of Volume 18 (Ocean Science) would like to acknowledge the following referees who have helped review the manuscripts published in this volume:

Vladimir Maderich
Hans Burchard
Y. Tony Song
Shunichi Koshimura
Hatta Mariko
Weifang Chen
Gwang Lee
Umberta Tinivella
Alexander Kurapov
Swapna Panickal

Jiang Zhu
Paul Harrison
Zhongming Lu
Ronghua Zhang
Martin Jakobsson
Boris Baranov
Xiaopei Lin
Xuejing Zhang
Youmin Tang

PREFACE

The present volume set of Advances in Geosciences (ADGEO) contains papers from the Busan annual meeting in 2008 and some from the Singapore annual meeting in 2009. As Editor-in-Chief, I must apologize to the AOGS members and authors for this delay. Since 2006 we have published 20 volumes in total. This publication project has been supported by the AOGS Council, World Scientific Publication Company (WSPC), the team of hard working editors and the broad membership and participants of AOGS. As with the main purpose of the Society, ADGEO is meant to promote information exchange and to forge scientific cooperation in the area of Earth science and environmental study. As witnessed by the negotiation efforts at the United Nations Climate Change Conference in Copenhagen in December 2009, all these issues have become more and more important and vital in the Asia-Pacific region. It is not a matter of exaggeration in saying that the solution to global warming, if there is one, has to come from the emerging economies and developing countries covered by AOGS. By design, ADGEO has its fundamental role to play. In practical terms, it is actually a difficult task because of many factors involved in deciding the quality of manuscripts, editorial and review processes, publication procedure, scientific impacts, readership, policy of the AOGS Council, and last but not least, marketing from the point of view of the publisher. Any small mishap in this long chain of interactive steps could lead to a major discontinuity. We have encountered such a situation with the publication of the Busan manuscripts. It is only with the cooperation of the authors, the ADGEO editorial team, WSPC, and the AOGS Secretariat Office, that we are able to produce these volumes, albeit a long delay. With this lesson learned, we hope to consolidate the ADGEO management and editorial system so that it would become an essential publication in our understanding of Earth and space science and information tool books in the battle against climate change. Finally, I would like to take this opportunity to thank the Volume Editors-in-Chief who are the driving force in making ADGEO possible: A. Bhardwaj (Planetary Science), M. Duldig (Solar and Terrestrial Science), J.P. Gan (Ocean Science), J.H. Oh (Atmospheric

Science), N.S. Park (Hydrological Science), K. Satake and C.H. Lo (Solid Earth). They have to work very hard to ensure both the quantity and quality of the published papers in ADGEO. Of equal importance, the support from WSPC is essential and its foresight in identifying the academic and social values of Earth science and environmental study to be sustained and articulated by ADGEO is very much appreciated.

Wing-Huen Ip
Editor-in-Chief

PREFACE TO OS VOLUME

Ocean Science is a branch of Geosciences and plays an important role in climate and environment variability. This volume collects papers that present recent developments and findings in Ocean Science by researchers from both the Asia Oceania region and other parts of the world. Here, a wide range of topics is covered, including ocean currents, waves, geophysical fluid dynamics, marine organisms, the ecosystem, biogeochemical substances and physical properties within the ocean, in the disciplines of physical, biological and chemical oceanography. The papers address important issues in climate and marine environment variability in both regional and global oceans. Results are obtained from analyses based on either cutting-edge numerical modeling techniques or field and laboratory measurements. The spectrum of topics covers ocean variability in spatial and temporal scales ranging from coastal ocean to basin scale.

Jianping Gan
OS Volume Editor-in-Chief

CONTENTS

Advances in Geosciences
Vol. 18: Ocean Science (2008)
Eds. Jianping Gan et al.

ELEVATED PHYTOPLANKTON BIOMASS IN MARGINAL SEAS IN THE LOW LATITUDE OCEAN: A CASE STUDY OF THE SOUTH CHINA SEA

KON-KEE LIU*

*Institute of Hydrological and Oceanic Sciences,
National Central University
Jhongli, Taiwan 320
kkliu@ncu.edu.tw*

CHUN-MAO TSENG

*Institute of Oceanography,
National Taiwan University
Taipei, Taiwan 106*

TZU-YING YEH

*Institute of Hydrological and Oceanic Sciences,
National Central University
Jhongli, Taiwan 320*

LI-WEN WANG

*Institute of Hydrological and Oceanic Sciences,
National Central University
Jhongli, Taiwan 320*

The sea surface chlorophyll concentrations in low latitude marginal seas, such as the South China Sea or Gulf of Mexico, are higher than those in the adjacent open oceans by >60%, the biomass higher by >28% and primary productivity higher by >100%. It is demonstrated by a coupled physical-biogeochemical model of the South China Sea that the enhanced phytoplankton growth in the SCS is probably mainly caused by rich supply of nutrients from localized upwelling along margins of the basin, whereas the river-borne nutrients may contribute to the enhancement to a relatively small degree. Although the upwelling process is wind-driven, it is the confinement of the marginal seas by the surrounding land masses that gives rise to the occurrences of upwelling under certain favorable conditions.

* Work supported by grant 97-2628-M008-001 of the National Science Council, Taiwan. This is NCU-IHOS contribution No. 95. This paper benefits from constructive comments of anonymous reviewers.

1. Introduction

It is noteworthy that the sea surface Chl-a concentration in the South China Sea (SCS) is on average about twice that in the adjacent West Philippine Sea (WPS) (Fig. 1). It has been widely observed that the integrated chlorophyll in the euphotic zone is positively related to the sea surface chlorophyll concentration.[1] The elevated chlorophyll concentration in the SCS indicates an elevated phytoplankton biomass. In addition, the primary production in the ocean can be successfully calculated from sea surface chlorophyll concentration in most area of the open ocean.[2] The elevated sea surface chlorophyll concentration in the SCS also implies elevated primary productivity.

Similar contrast in sea surface chlorophyll concentration also occurs to the Gulf of Mexico and the adjacent western North Atlantic Ocean with the Gulf of Mexico having a higher Chl-a concentration (Fig. 2). Both pairs comprise a marginal sea and an oligotrophic open ocean in the same tropical–subtropical zone; in each pair the phytoplankton standing stock is

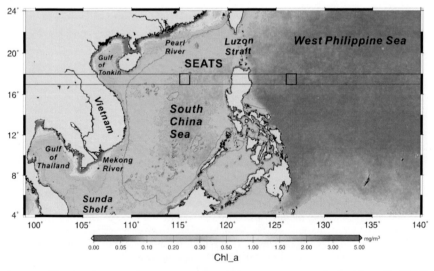

Fig. 1.   The distribution of mean sea surface Chl-a concentration derived from SeaWiFS ocean color data averaged over the period from September 1997 to December 2003. The contours in red indicate the 200 m isobaths. The black lines indicate the zone, where the zonal variation of Chl-a concentration is presented in Fig. 3(a). The two black boxes indicate the region, where the wind data are presented in Fig. 5(a). The red dot indicates the SEATS Station.

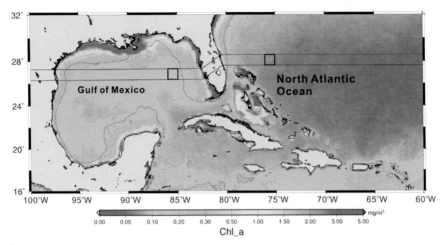

Fig. 2.   The same as Fig. 1 except for the Gulf of Mexico and the adjacent area. The black lines indicate the zone, where the zonal variation of Chl-a concentration is presented in Fig. 3(b). The two black boxes indicate the region, where the wind data are presented in Fig. 5(b).

enhanced in the marginal sea in comparison with the open ocean. Such a contrast has led to the inclusion of marginal seas, which are defined as semi-enclosed seas adjacent to continents, in the coastal ocean or continental margin systems.[3,4] This means that the marginal seas are more akin to the coastal ocean than to the open ocean. It is worth discussing why the marginal sea and the adjacent open ocean in each pair exhibit different biogeochemical characteristics, though they are subject to similar physical forcing, such as wind and solar irradiance.

In this paper, we intend to present quantitative evidence of elevated phytoplankton biomass in two of the largest low latitude marginal seas and to explore processes that may be responsible for the different biogeochemical behavior in the marginal sea as opposed to the open ocean. One obvious reason for the difference is the supply of nutrients from rivers in marginal seas.[5] It has also been suggested that nutrient pumping from the underlying water body in the marginal seas is mainly responsible for the elevated chlorophyll concentration.[6,7] Because the vertical transport of nutrient is difficult to quantify by observation, we use a coupled physical-biogeochemical model of the South China Sea to estimate the vertical nutrient flux, which may be compared to the river loads of nutrients. The model may also provide insight into the biogeochemical responses of the South China Sea to seasonal pulses of nutrient supply.

2. Materials and Methods

The shipboard measurements used in this study were obtained from the South-East Asian Time-series Study (SEATS).[8] The SEATS station is located at 18°N and 116°E (Fig. 1). Discrete seawater samples were collected with depth in GO-FLO bottles that were mounted onto a Rosette sampling assembly (General Oceanic). Separate sub-samples were filtered onboard the research vessel. The filters were stored at −20°C and then returned to the shore-based laboratory for the fluorometric determination of chlorophyll-a (Chl-a).[9]

The satellite obtained sea surface chlorophyll-a (Chl-a) values used in this study are a SeaWiFS data product from the following web site: (http://daac.gsfc.nasa.gov/data/dataset/SEAWIFS/01_Data_Products/index.html) The data presented are mean values averaged for each pixel over the period between September 1997 and December 2003. The daily wind speed data presented were provided by NCEP Re-analysis at http://www.cdc.noaa.gov/cdc/data.ncep.reanalysis2.html#sktnote.

An improved three-dimensional coupled physical-biogeochemical model has been developed for the SCS.[10] The model has a horizontal resolution of 0.4° in the domain 2–24.8°N and 99–124.6°E and 21 layers in the vertical. The nitrogen-based biogeochemical model has four compartments, dissolved inorganic nitrogen (DIN), phytoplankton (Phy), zooplankton (Zop) and detritus (Det) with a variable chlorophyll-to-nitrogen ratio (R) in phytoplankton. The upper and lower bounds of R are set to 3.5 and $1.0 \, \mathrm{g \, mol \, N^{-1}}$, respectively, which correspond to organic carbon-to-chlorophyll ratio of 23 and 80, respectively.[10] The model includes the benthic-pelagic coupling for nitrogen. The particulate nitrogen in detritus that hits the bottom is transformed into DIN with a 14% removal by benthic denitrification on the seafloor.[10,11]

The initial conditions of the model are the January temperature and salinity fields of climatological hydrography[12] and the mean DIN field.[13] The model is driven by climatological monthly mean winds,[14] monthly sea surface temperature, and seasonal surface salinity.[12] The flow field in the third year of the model run shows little changes from those in the previous years and the biogeochemical tracers also reach quasi-steady state.[15,16]

Therefore, the output from the third year of the model run is presented in all cases. The model output is compared with observations from SEATS.

3. Results and Discussion

We first present quantitative evidence showing the elevated phytoplankton biomass in the SCS and the Gulf of Mexico. Then we discuss possible mechanisms responsible for the enhancement with the help of numerical modeling for the SCS.

3.1. *Sea surface chlorophyll and phytoplankton biomass*

The zonal variations of sea surface Chl-a concentration are used to demonstrate the contrast between the marginal sea and the adjacent open ocean in the low latitudes (Fig. 3). The case of the SCS-WPS pair is illustrated by the change of Chl-a along latitude 17.5°N from 100°E to 140°E (Fig. 3(a)). The value at each longitude is the average Chl-a value from the zonal band, 17–18°N, shown in Fig. 1(a). The Chl-a values are high in coastal zones, where elevated nutrient levels due to terrestrial input may enhance phytoplankton growth. However, suspended sediments and colored dissolved organic matter from land in the coastal zone may cause overestimation of Chl-a concentration derived from ocean color data. Away from the coastal zone the Chl-a value drops precipitatiously. In regions with water depth greater than 200 m and 50 km or farther from the coast inside the SCS, the Chl-a concentrations are rather low. In the deeper region away from the coast, the sea surface Chl-a concentration ranges between 0.09 and $0.13\,\mathrm{mg\,m^{-3}}$ with a mean value of $0.10\,\mathrm{mg\,m^{-3}}$. In the eastern sea board of the Philippines, the Chl-a concentration is only about $0.08\,\mathrm{mg\,m^{-3}}$ at a distance 50 km to the east coast of the Luzon Island and decreases to $0.06\,\mathrm{mg\,m^{-3}}$ further east.

In the deeper part of the Gulf of Mexico away from the coast (Fig. 3(b)), the sea surface Chl-a concentration ranges between 0.11 and $0.15\,\mathrm{mg\,m^{-3}}$ with a mean value of $0.12\,\mathrm{mg\,m^{-3}}$. In the adjacent North Atlantic Ocean, the sea surface Chl-a concentration is $0.1\,\mathrm{mg\,m^{-3}}$ at about 50 km from the Florida coast and decreases to $0.05\,\mathrm{mg\,m^{-3}}$ or lower further to the east. It is clear that in the interior of the two marginal seas, where the water depths attain 3000 m or greater, the sea surface Chl-a concentrations are about twice as high as those in the adjacent open ocean.

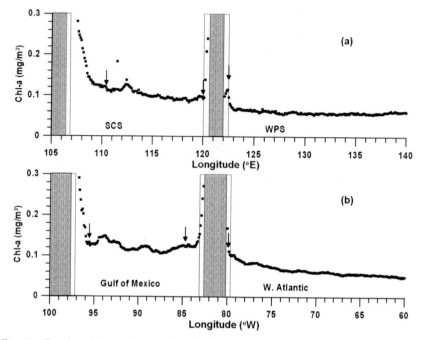

Fig. 3. Zonal variation of sea surface Chl-a concentration from the marginal sea to the adjacent open ocean. The contrast is clear that the Chl-a concentration within the marginal sea is considerably higher. The shaded areas indicate land masses. The open boxes indicate the coastal zone within 50 km from the coast. The arrows indicate the positions of the 200 m isobaths. (a) For the SCS and the adjacent WPS. The values are averaged over the band from 17–18°N. (b) For Gulf of Mexico to the adjacent North Atlantic Ocean. The values are averaged over the band of 26.3–27.3°N within the Gulf, and 27.7–28.7°N to the east of Florida.

Based on observations at the SEATS Station (Fig. 4), the integrated chlorophyll, I-chl $(mg\,m^{-2})$, in the euphotic zone, which may represent the phytoplankton biomass, is related to the surface chlorophyll, S-chl $(mg\,m^{-3})$, for the SCS by the following equation.

$$\text{I-chl} = 54.5(\text{S-chl})^{0.471}, \quad R^2 = 0.768. \tag{1}$$

The relationship resembles that found by Morel and Berthon,[1] who reported a similar exponent (0.425) but a smaller coefficient (38). The relationship found at the SEATS station appears to fit observations from the Gulf of Mexico.[17] If this relationship also holds true for the WPS, the integrated chlorophyll in the SCS proper is on average 28% higher than that in the adjacent WPS. Similarly, the integrated chlorophyll in the Gulf of Mexico

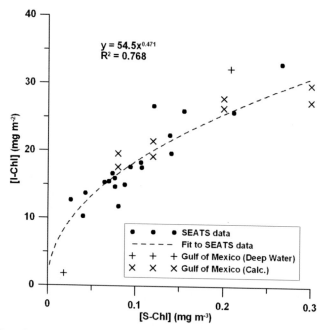

Fig. 4. The relationship between the integrated chlorophyll (I-chl) in the euphotic zone and the surface chlorophyll (S-chl) observed at the SEATS Station. Also plotted are two observations from deep water stations in the Gulf of Mexico and typical data calculated from relationships between S-chl and I-chl derived from 352 observed profiles in the Gulf of Mexico.[17]

interior is on average 35% higher than that in the adjacent North Atlantic Ocean.

Because of the more abundant light in the water column near the surface, the phytoplankton in the surface layer contributes more to primary production than the subsurface biomass. The average primary production (PP) in the SCS proper was found to be $390 \, \mathrm{mg \, m^{-2} \, d^{-1}}$ for summer and $546 \, \mathrm{mg \, m^{-2} \, d^{-1}}$ for winter,[18] whereas the mean PP in the adjacent WPS was found to be around $200 \, \mathrm{mg \, m^{-2} \, d^{-1}}$ or lower.[2] The enhanced PP in the SCS is more than double as compared to that in the WPS. Especially important is the higher winter production under the stronger winter monsoon. This is to be addressed in the following.

3.2. *Wind speed*

Because wind mixing is often the dominant factor controlling primary productivity in the ocean,[19] it is desirable to check whether changes in the

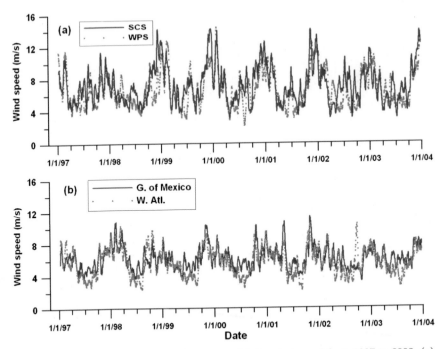

Fig. 5.   Time-series of 15-day running mean of daily wind speed from 1997 to 2003. (a) For the SCS and WPS regions indicated by the two black boxes shown in Fig. 1. (b) For the Gulf of Mexico and adjacent North Atlantic Ocean regions indicated by the two black boxes shown in Fig. 2.

wind strength are responsible for the contrast. The time-series of wind speed in different regions are compared in Fig. 5. For the SCS-WPS pair, the daily wind speeds in the region within 17–18°N and 115–116°E (near the SEATS Station) are compared to those in the region within 17–18°N and 126–127°E (Fig. 5(a)). The winter-summer difference was clear in the annual cycles. The northeast monsoons in winter were considerably stronger than the southwest monsoons in summer, but the latter still show secondary peaks in mid-years. As a result, weakest winds occurred during inter-monsoon periods. Over the period from January 1997 to December 2003, the wind speeds in the adjacent regions were quite similar. The mean values were 7.2 ± 3.5 and $6.7 \pm 3.4 \, \mathrm{m\,s^{-1}}$, respectively, for the SCS and WPS.

The winter-summer contrast was also evident for the Gulf of Mexico region (Fig. 5(b)). Lacking the monsoonal system, the amplitude of the winter-summer variation was not as large as that found in the SCS region. Unlike the monsoonal system, secondary peaks of wind speed in summer were mostly absent in the Gulf of Mexico. The mean values were 6.1 ± 2.7 and $5.7 \pm 2.9\,\mathrm{m\,s^{-1}}$, respectively, for the Gulf of Mexico and the adjacent open ocean.

Although it cannot be ruled out that the slightly stronger wind speed in the marginal seas may contribute to the more enhanced phytoplankton growth, it is unlikely that the small differences in wind speed are solely responsible for the elevated phytoplankton biomass and the much enhanced primary productivity within the marginal sea as compared to the adjacent open ocean. Using the SCS as a case study, we explore the processes that may contribute to the contrast in greater detail.

3.3. *Monsoon-driven variation of chlorophyll*

The effect of wind-driven variation of chlorophyll in the SCS can be examined with the coupled physical-biogeochemical model of Liu et al.[10] The model-generated seasonal variation of sea surface chlorophyll concentration in the vicinity of the SEATS Station is compared to the observations obtained between September 1999 and October 2003 (Fig. 6). There is good agreement between the model output and observations over the annual cycle. The model successfully reproduced the winter-summer contrast with the lowest chlorophyll concentration $(0.06\text{--}0.09\,\mathrm{mg\,m^{-3}})$ occurring from April to September, while the observed values ranged from 0.03 to $0.09\,\mathrm{mg\,m^{-3}}$. The model results showed high values in the months

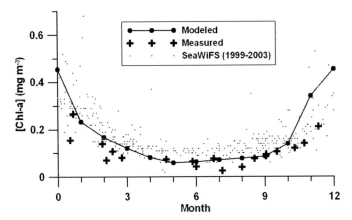

Fig. 6.   Seasonal variation of modeled and observed sea surface Chl-a concentration at the SEATS Station.

under the strong northeast monsoon, namely, from November to February, mainly due to enhanced upwelling in the northern SCS near Luzon and vertical mixing.[16] The modeled values ranged from 0.17 to 0.46 mg m^{-3}, while the observed values ranged between 0.12 and 0.27 mg m^{-3}. The model appeared to overestimate the chlorophyll concentrations slightly as compared to observations. The Chl-a values derived from SeaWiFS data in the vicinity ($1° \times 1°$) of the SEATS station from 1999 to 2003 are also shown for comparison (Fig. 6). The SeaWiFS values show a seasonal trend very similar to those manifested by observed and modeled values, but are somewhat higher in magnitudes compared to observations.[8] This is consistent with the estimated accuracy of $\pm 35\%$ with a relatively small bias for SeaWiFS Chl-a data.[20] Nevertheless, they confirm occurrences of elevated Chl-a values in northern SCS in winter, when no observed values were obtained.

The modeled spatial distributions of chlorophyll in different seasons are compared to images of chlorophyll concentration derived from the long-term averaged monthly SeaWiFS data (Figs. 7–10). The monthly model results are represented by the model output for the 15th of each month. In April, before the onset of the southwest monsoon, both the SeaWiFS image (Fig. 7(a)) and model result (Fig. 7(b)) show very low chlorophyll concentration everywhere in the SCS except in the coastal zone.

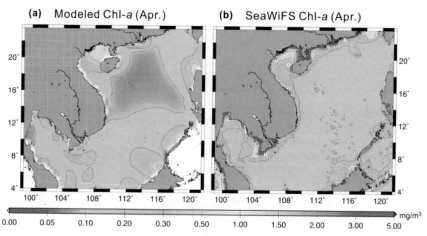

Fig. 7. Sea surface Chl-a distribution in the South China Sea for April. The contour levels below 1 mg m^{-3} are 0.1, 0.2 and 0.5 mg m^{-3}. (a) Model output generated by the improved biogeochemical model of Liu et al.[10] (b) The long-term average of monthly SeaWiFS chlorophyll.

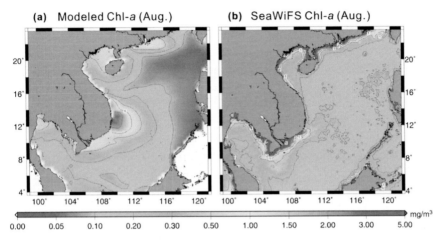

Fig. 8. The same as Fig. 7 except for August.

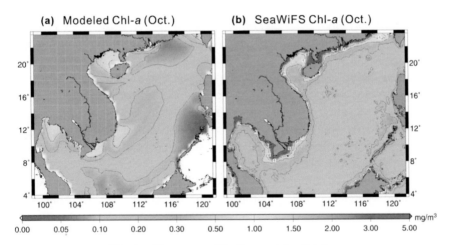

Fig. 9. The same as Fig. 7 except for October.

Under the summer monsoon (August), the model produces relatively high chlorophyll concentrations ($1 \, \text{mg m}^{-3}$ or higher) off the Pearl River Mouth, in the Gulfs of Tonkin and Thailand, and off the east coast of Vietnam and the Mekong River Mouth (Fig. 8(a)), while chlorophyll concentrations elsewhere are mostly below $0.2 \, \text{mg m}^{-3}$. The SeaWiFS chlorophyll image (Fig. 8(b)) shows a band of intense chlorophyll concentration of about 100 km wide extending from the Mekong River mouth towards the northeast. Although the Mekong runoff may induce high productivity, the extensive high chlorophyll patch far from the river mouth is most likely induced by the upwelling under the southwest monsoon.[21] The modeled chlorophyll distribution off the east coast of Vietnam appears too much enhanced, probably resulting from the coarse model resolution and insufficient advection.[16] On the other hand, the modeled chlorophyll concentrations in the coastal zone are not as high as the SeaWiFS image shows. It is cautioned that some of the very high SeaWiFS chlorophyll concentrations ($>5\,\mathrm{mg\,m^{-3}}$) near the coast may be caused by colored dissolved organic matter or suspended sediment in the Case II waters.[22]

The reasonably good simulation of chlorophyll distribution in the Gulf of Thailand is much improved compared to the original model;[16] this is mainly due to the inclusion of a more realistic benthic boundary condition.[10] In contrast to the high chlorophyll in the western half of the SCS, both the model and the satellite image show very low chlorophyll concentrations on the eastern half of the basin probably due to downwelling along the eastern boundary of the basin.[21]

The chlorophyll concentrations drops to very low levels in October (Fig. 9), when the summer monsoon subsides and the winter monsoon has yet to pick up. This is similar to the condition during the spring inter-monsoon period (Fig. 7). A notable patch of slightly elevated chlorophyll concentration that extends from the coastal zone of Vietnam towards northeast (Fig. 9(a)) appears to be a remnant of the summer upwelling. The SeaWiFS image (Fig. 9(b)) shows a similar distribution pattern and range of Chl-a concentration, but the remnant of summer upwelling is absent and the coastal features are more confined.

As the winter monsoon becomes stronger in December (Fig. 5(a)), the model predicts high chlorophyll concentrations (up to $2\,\mathrm{mg\,m^{-3}}$) off northwest Luzon. The high chlorophyll patch extends southwestward forming a band of moderately elevated chlorophyll concentrations (up to $1\,\mathrm{mg\,m^{-3}}$), reaching the Sunda Shelf region in the southern end of the SCS (Fig. 10(a)). The areas off northwestern Luzon and Sunda shelf are regions of upwelling according to model-predicted circulation.[21] The SeaWiFS image (Fig. 10(b)) shows regions of elevated chlorophyll concentration off northwest Luzon and to the south of Vietnam. However, the enhancement is not as strong as the model predicts.

As shown above, the modeled features of sea surface Chl-a distribution in the SCS show qualitative agreement with the SeaWiFS images, but the details differ considerably. The main reason is the coarse resolution of the model, which does not allow occurrences of mesoscale eddies,[16]

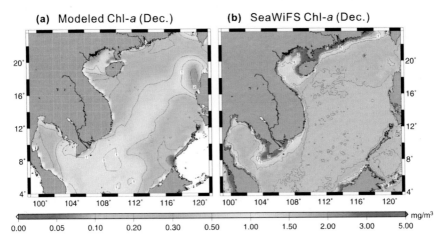

Fig. 10.   The same as Fig. 7 except for December.

while these eddies have proven to be a very important feature in the SCS.[23] Improvement of the present model by adopting a finer horizontal resolution is necessary to better simulate the spatial distribution of Chl-a. Although the modeled Chl-a distributions do not match the SeaWiFS data precisely, the modeled seasonal variation of primary production in the SCS matches observations reasonably well.[10] The modeled annual mean primary production is $499 \, \mathrm{mg \, C \, m^{-2} \, d^{-1}}$, which compares well with the reported mean values of $390 \, \mathrm{mg \, C \, m^{-2} \, d^{-1}}$ and $546 \, \mathrm{mg \, C \, m^{-2} \, d^{-1}}$ for summer and winter, respectively.[10,18]

3.4. *Nutrient demand and supply*

Since the coupled model for the SCS has done a reasonably good job in simulating phytoplankton growth in the SCS, we use it to provide information on the demands and supplies of nutrients for phytoplankton growth (Table 1). The nutrient demands can be calculated from the modeled primary production under the assumption of C/N/P ratio of 106:16:1.[24] The required nutrient loads are calculated for the entire SCS with an area of 3.35 million $\mathrm{km^2}$ and listed in Table 1.

The main supply of nutrients to the upper water column of the SCS defined as the top 100 m is adopted from the estimates of Liu *et al.*[25] and presented in Table 1. Because the upper bound of observed euphotic zone depths in the SCS was close to 100 m,[26] the nutrient supply in the top 100 m is crucial to phytoplankton growth. It is clear that upwelling

Table 1. Nutrient demands and supplies in the upper water column (0–100 m) in the SCS.

	Category	N (Gmol y^{-1})	N-flux (μmol $m^{-2} d^{-1}$)	P (Gmol y^{-1})	P-flux (μmol $m^{-2} d^{-1}$)
Nutrient demand	Primary production	6282	5137	393	321
	Upwelling*	1893	1548	127	104
	Precipitation*	23	19	1.0	0.8
Nutrient supply	Runoff*	64	52	0.8	0.7
	Nitrogen fixation	127	104		
	Typhoon-driven mixing*	40	33	3.0	2.5
	Internal wave breaking*	23	19	1.7	1.4

*Note: Only dissolved inorganic nutrients are considered.

of the subsurface water provides the largest nutrient flux to the upper water column. The alternating monsoons drive upwelling in the SCS,[21,27,28] notably in three regions: off northwestern Luzon and around the shelf break in northern Sunda Shelf in winter and off the central Vietnamese coast in summer. It is the localized upwelling that brings nutrients from the deep to the upper water column.[21]

The river runoff provides only about 1% of the nitrogen demand for primary production and about 0.2% of phosphorus demand. It is noted that only the dissolved inorganic nutrients, which are most readily available to phytoplankton uptake, are considered here. Although river-borne nutrients are not important to the SCS as a whole, they have strong impacts on the coastal zone near the river mouth. Besides, they are ultimately important in a long run because they are the major sources to the nutrient reserve in the ocean.

Typhoon-driven mixing[29] and internal wave breaking[30] are two special processes that contribute to nutrient pumping in the SCS. The nutrient fluxes brought by precipitation originate from Aeolian transports of nutrients from land. In the SCS region, biomass burning could be an important source of nutrient fallout.[31]

It is worth mentioning that the nutrient supplies presented are only the external sources to the euphotic zone, while the internal source, namely, the regenerated nutrient supply within the euphotic zone is not included.

Consequently, the sum of the external nutrient supply does not match the nutrient demands for primary production but accounts for only a fraction of the total demands. The rest of the demands come from the recycled nutrients.

4. Final Remarks

The elevated phytoplankton biomass in the SCS and the much enhanced primary production is probably mainly caused by a rich supply of nutrients from localized upwelling along margins of the basin,[21] while river-borne nutrients may contribute to the enhancement to a small degree. It is the upwelling process that keeps a steady supply of nutrients to the upper water column, while the wind mixing in the surface layer is only the last leg in the chain of nutrient supply processes.

The localized upwelling mostly occurs along the continental slope or in the water body close to it. In the SCS, the upwelling off NW Luzon is attributed to subsurface convergence of the eastern boundary current.[27] The positive wind stress curl, which is likely a result of atmosphere-land mass interaction also induces Ekman suction or upwelling.[32] As the southward current along the western shore of the SCS impinges on the Sunda Shelf in winter, the subsurface water is uplifted by the shoaling topography, inducing upwelling. The upwelling off Vietnam is induced by the separation of the western boundary jet from the Vietnamese coast in summer. The seaward drift causes an upwelling center off Vietnam at about 12°N. These processes that occur because of the confinement of the marginal sea or the land-mass surrounding it are likely the main reason for the enhanced phytoplankton growth. Further studies using more sophisticated modeling approaches in wider domains are warranted to further our understanding of these critical issues.

References

1. A. Morel and J. F. Berthon, *Limnol. Oceanogr.* **34** (1989) 687.
2. M. J. Behrenfeld and P. G. Falkowski, *Limnol. Oceanogr.* **42** (1997) 1.
3. A. Robinson and K. H. Brink, eds., *The Sea, vol. 14A,B, The Global Coastal Ocean: Interdisciplinary Regional Studies and Synthesis* (Harvard University Press, Cambridge, 2006).
4. K.-K. Liu, L. Atkinson, R. Quiñones and L. Talaue-McManus, eds., *Carbon and Nutrient Fluxes in Continental Margins: A Global Synthesis* (Springer, Berlin, 2009).

5. S. V. Smith, D. P. Swaney, L. Talaue-McManus, J. D. Bartley, P. T. Sandhei, C. J. McLaughlin, V. C. Dupra, C. J. Crossland, R. W. Buddemeier, B. A. Maxwell and F. Wulff, *Bioscience* **53** (2003) 235.
6. R. Wollast, in *The Sea, vol. 10, The Global Coastal Ocean: Processes and Methods*, eds. A. Robinson, K. H. Brink (Wiley, New York, 1998), pp. 213–252.
7. C.-T. A. Chen, K.-K. Liu and R. W. MacDonald, in *Ocean Biogeochemistry: The Role of the Ocean Carbon Cycle in Global Change*, ed. M. J. R. Fasham (Springer, Berlin, 2003), pp. 53–97.
8. C. M. Tseng, G. T. F. Wong, Lin, II, C. R. Wu and K. K. Liu, *Geophys. Res. Lett.* **32** (2005) L08608, doi:10.1029/2004GL022111.
9. J. D. H. Strickland and T. R. Parsons, *A Practical Handbook of Seawater Analysis* (Fisheries Research Board of Canada, Ottawa, 1972), p. 310.
10. K.-K. Liu, Y.-J. Chen, C.-M. Tseng, I.-I. Lin, H. Liu and A. Snidvongs, *Deep-Sea Res. II* **54** (2007) 1546.
11. K. Fennel, J. Wilkin, J. Levin, J. Moisan, J. O'Reilly and D. Haidvogel, *Global Biogeochem. Cycles* **20** (2006) GB3007, doi:10.1029/2005GB002456.
12. S. Levitus, Climatological atlas of the world ocean, NOAA Professional paper No. 13, U.S. Government Printing Office (1982).
13. M. E. Conkright, S. Levitus and T. P. Boyer, World Ocean Atlas 1994, vol. 1: Nutrients, U.S. Government Printing Office (1994).
14. S. Hellerman and M. Rosenstein, *J. Phys. Oceanogr.* **13** (1983) 1093.
15. P. T. Shaw and S. Y. Chao, *Deep-Sea Res. I* **41** (1994) 1663.
16. K. K. Liu, S. Y. Chao, P. T. Shaw, G. C. Gong, C. C. Chen and T. Y. Tang, *Deep-Sea Res. I* **49** (2002) 1387.
17. R. M. Hidalgo-Gonzalez and S. Alvarez-Borrego, *Ciencias Marinas* **34** (2008) 197.
18. X. Ning, F. Chai, H. Xue, Y. Cai, C. Liu, G. Zhu and J. Shi, *J. Geophys. Res. C. Oceans* **109** (2004) C10005, doi:10.1029/2004JC002365.
19. E. Sakshaug, K. Tangen and D. Slagstad, in *The Changing Ocean Carbon Cycle: A Midterm Synthesis of the Joint Global Ocean Flux Study*, eds. R. B. Hanson, H. W. Ducklow and J. G. Field (Cambridge University Press, Cambridge, 2000), pp. 19–36.
20. C. R. McClain, G. C. Feldman and S. B. Hooker, *Deep-Sea Res. II* **51** (2004) 5.
21. S. Y. Chao, P. T. Shaw and S. Y. Wu, *Deep-Sea Res. I*, vol. 43 (1996).
22. S. Sathyendranath, Remote sensing of ocean color in coastal and other optically-complex waters, Technical Report No. 3, IOCCG (2000).
23. C. R. Wu and T. L. Chiang, *Deep-Sea Res. II* **54** (2007) 1575.
24. A. C. Redfield, B. H. Ketchum and F. A. Richards, in *The Sea, vol. 2*, ed. M. N. Hill (Interscience Publishers, New York, 1963), pp. 26–77.
25. K.-K. Liu, C.-M. Tseng, C.-R. Wu and I.-I. Lin, in *Carbon and Nutrient Fluxes in Continental Margins: A Global Synthesis*, eds. K. K. Liu, L. Atkinson, R. Quiñones and L. Talaue-McManus (Springer, Berlin, 2009).
26. Y. L. L. Chen, *Deep-Sea Res. I* **52** (2005) 319.

27. P. T. Shaw, S. Y. Chao, K. K. Liu, S. C. Pai and C. T. Liu, *J. Geophys. Res. (Oceans)* **101** (1996) 16435.

28. M. J. B. Udarbe-Walker and C. L. Villanoy, *Deep-Sea Research Part I — Oceanographic Research Papers* **48** (2001) 1499.

29. I.-I. Lin, W. T. Liu, C. C. Wu, G. T. F. Wong, C. M. Hu, Z. Q. Chen, W. D. Liang, Y. Yang and K. K. Liu, *Geophys. Res. Lett.* vol. 30 (2003).

30. M.-K. Hsu, A. K. Liu and C. Liu, *Cont. Shelf Res.* **20** (2000) 389.

31. I.-I. Lin, J. P. Chen, G. T. F. Wong, C. W. Huang and C. C. Lien, *Deep-Sea Res. II* **54** (2007) 1589.

32. S. Y. Chao, P. T. Shaw and S. Y. Wu, *Prog. Oceanogr.* **38** (1996) 51.

Advances in Geosciences
Vol. 18: Ocean Science (2008)
Eds. Jianping Gan et al.
© World Scientific Publishing Company

NUMERICAL SIMULATION OF TROPICAL CYCLONE INTENSITY USING AN AIR-SEA-WAVE COUPLED PREDICTION SYSTEM

LIAN XIE[†,§,*], BIN LIU[†], HUIQING LIU[†], and CHANGLONG GUAN[‡]

[†] Department of Marine, Earth and Atmospheric Sciences,
North Carolina State University, Raleigh, NC 27695, USA
[‡] Physical Oceanography Laboratory, College
of Physical and Environmental Oceanography,
Ocean University of China, Qingdao, Shandong 266100, China
[§] College of Environmental Science and Engineering,
Ocean University of China, Qingdao, Shandong 266100, China
*xie@ncsu.edu

A coupled regional ocean and weather numerical modeling system (CROWN) is constructed and applied to simulate the interaction between the ocean and tropical cyclones. The CROWN modeling system consists of the regional version of the Weather Research and Forecasting (WRF) model; the Princeton Ocean Model (POM); and a third-generation ocean surface wave model (Simulating WAves Nearshore, SWAN or WAVEWATCH III, WW3). WRF is coupled to both POM and SWAN through the Model Coupling Toolkit (MCT). Sea surface temperature from POM, wave parameters from SWAN or WW3, and meteorological parameters from WRF are exchanged to compute air-sea momentum, sensible and latent heat fluxes. Air-sea fluxes caused by the effect of sea spray are also computed and added to the total air-sea fluxes, which are then passed to all three models. The ocean circulation model and the wave model are coupled not only through wave and sea spray induced surface fluxes, but also by exchanging radiation stress and bottom stress. The modeling system has been tested for both stand-alone and coupled simulations of Hurricane Katrina, as well as oceanic responses in terms of surface waves and storm surge. Coupled simulations reveal significant contributions from wind-wave and wave-current coupling, as well as sea spray effects to the intensity change of Hurricane Katrina as well as the storm surge and waves induced by the storm.

1. Introduction

The atmosphere and the ocean form an inextricably coupled fluid system on the surface of the earth. For instance, oceanic processes regulate the atmospheric climate and weather systems, which, in turn, affect the underlying ocean currents, waves, temperature and salinity, as well as the marine ecosystem. Thus, developing air-sea coupled modeling systems capable of predicting atmospheric processes at regional to global scales as well as the underlying oceanic circulation and waves provides a foundation for an integrated approach to numerical predictions of the coupled atmosphere–ocean system.

Coupled atmosphere–ocean models are widely used to study and predict climate variability and climate change, but only until recently, two-way coupling between the ocean and the atmosphere began to be adopted in regional weather and ocean forecasting, e.g. Xie[1] and the references therein. Early applications of regional coupled models did not consider ocean surface waves as an explicit component of the air-sea coupled system, nor did they consider time-dependent land-water boundaries which may change in time due to inundation and drying caused by storms and sea-level rise. Xie et al.[2] developed a wave–current coupled system based on a three-dimensional coastal ocean circulation model (the Princeton Ocean Model, POM) and a third generation wave model (WAM). A limitation of the coupled system is that the radiation stress was computed according to a vertically uniform, depth-averaged wave model, whereas the three-dimensional POM requires the input of a radiation stress that varies with water depth.[3,4] Wave–current coupled models that include vertically-varying radiation stress have been developed in recent years by introducing a vertical radiation stress profile.[4,5] Xie et al.[6] developed a wave–current coupled modeling system which incorporates a depth-dependent radiation stress, surface and bottom shear stresses, as well as inundation and drying.

In addition to wave–current coupling, interactions between wind and ocean waves are also a critical component of an integrated air-sea coupled modeling system. In climate models, air-sea interactions often focus on the effect of wind stress on the ocean and the thermodynamic feedback from the ocean to the atmosphere through sensible and latent heat fluxes based on either constant or wind speed dependent heat and moisture transfer coefficients. It is well known that air-sea heat, moisture and momentum exchange processes depend strongly on the properties of ocean surface waves, which is, in turn, a function of surface wind.[7,8] Recent studies further indicate that air-sea exchange coefficients are not a linear function of wind speed at high wind speed.[9] In fact, observations show that when wind speed exceeds a critical value of approximately 25–$35\,\mathrm{m\,s^{-1}}$, air-sea transfer coefficients do not increase with wind speed, but are either decreasing or tapering off as wind speed increases further.[9] Thus, a properly developed air-sea coupled modeling system must have the capability to reproduce the observed relationship between wind speed and sea state.

Although extensive studies on wind–wave interaction, air-sea interaction, and wave–current interaction have been carried out, wind–current–wave three-way interactions have not yet been fully implemented in air-sea coupled models. The development and implementation of such a three-way coupled atmosphere–ocean modeling and prediction system is the focus of this study.

2. Description of the Coupled Regional Ocean and Weather Numerical Modeling System (CROWN)

The CROWN modeling system consists of the regional version of the Weather Research and Forecasting (WRF) model;[10] a regional ocean circulation model (the Princeton Ocean Model, POM); and a third-generation ocean surface wave model, the SWAN (Simulating WAves Nearshore) model[11] or the WAVEWATCH III (WW3) model.[12]

The coupling between the components of the atmosphere and the ocean is carried out by using the Model Coupling Toolkit (MCT).[13,14] MCT is a set of open source software tools written in Fortran90 and works with the MPI communication protocol. It allows efficient data transfer between the model components, and provides interpolation algorithms for the transferred variables as well. The MCT has been used as the base for the Community Climate System Model coupler,[15] and to couple the Regional Ocean Modeling System (ROMS) with other models.[16]

In the coupling processes of the CROWN modeling system, sea surface temperature from POM, wave parameters from SWAN or WW3, meteorological parameters from WRF are exchanged among the model components to computes air-sea momentum, sensible and latent heat fluxes. Figure 1 is a schematic illustration of the coupled wind–wave–current system. The atmospheric model drives the ocean model and the wave model through surface forcing (surface momentum and heat fluxes), and provides the environment variables such as temperature (T), specific humidity (q), surface pressure (p), and frictional velocity u_* for the estimation of sea spray heat fluxes. The wave model provides wave parameters for determining wave state, which in turn influences sea surface roughness parameters

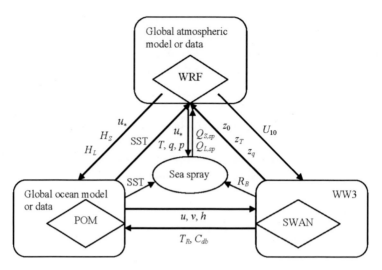

Fig. 1.   Schematic depiction of atmosphere-wave-ocean coupling.

(z_0, z_T and z_q) as well as the spray heat fluxes ($Q_{S,sp}$ and $Q_{L,sp}$). The ocean model provides sea surface temperature (SST) to the coupled system. Also, wave-current interaction is considered by coupling the ocean circulation model to the wave model through exchanging radiation stress and bottom stress. In addition, the dissipative heating at the surface layer in the atmospheric model is also taken into account. The details of each component of the modeling system, the parametrization of air-sea interaction, and the coupling procedure are described in the following sections.

2.1. The atmospheric model

The Weather Research and Forecasting (WRF) Model with the Advanced Research WRF (ARW) core[10] is used as the atmospheric component in the coupled system. The WRF is a fully compressible, non-hydrostatic mesoscale numerical weather prediction model suitable for a broad spectrum of applications across scales ranging from meters to thousands of kilometers. It has been developed via a collaborative partnership, principally among the National Center for Atmospheric Research (NCAR), the National Oceanic and Atmospheric Administration (the National Centers for Environmental Prediction (NCEP) and the Global System Division (GSD), formal Forecast Systems Laboratory (FSL), of Earth System Research Laboratory (ESRL), the Air Force Weather Agency

(AFWA), the Naval Research Laboratory, Oklahoma University, and the Federal Aviation Administration (FAA). The model uses terrain-following hydrostatic pressure coordinate system in vertical dimension. The grid staggering is the Arakawa C-grid. The Runge–Kutta 2nd and 3rd order time integration schemes and 2nd to 6th order advection schemes are used in both horizontal and vertical directions. The WRF model incorporates physical processes including microphysics, cumulus parametrization, planetary boundary layer, surface layer, land-surface, and long-wave and shortwave radiations, with several options available for each process. More detailed description of the WRF model is referred to Skamarock *et al.*[10] and Wang *et al.*[17]

2.2. *The wave model*

The wave model component of the coupled atmosphere system is a third generation model with a choice of WW3 or SWAN. The governing equation of a third generation wave model is the action balance equation, which can be written as:[18]

$$
\frac{\partial N}{\partial t} + \nabla_{\vec{x}} \cdot [(\vec{c}_g + \vec{U})N] + \frac{\partial c_\sigma N}{\partial \sigma} + \frac{\partial c_\theta N}{\partial \theta} = \frac{S_{tot}}{\sigma}, \tag{1}
$$

where σ is the relative (intrinsic) frequency, N is wave action density equal to energy density divided by relative frequency, $\vec{c}_g$ is the wave action propagation speed in $\vec{x}$-space, c_σ and c_θ are the propagation velocities in spectral space (σ, θ). The right hand side (S_{tot}) is the total of source or sink terms expressed as wave energy density. In deep water, S_{tot} is dominated by three terms including wind input, quadruplet wave-wave interactions, and dissipation. While in shallow water, triad wave interactions and depth-induced wave breaking should also be considered. WW3 tends to be more efficient at large scale simulation, whereas SWAN has some advantages at small scale simulation, especially when concerning shallow water conditions.

2.3. *The ocean circulation model*

The regional ocean circulation model selected for the coupled system is the Princeton Ocean Model (POM) which is originally developed by Mellor and Blumberg.[19] POM is a three-dimensional, primitive-equation model that uses a sigma coordinate in the vertical, an "Arakawa C" grid scheme in the horizontal, and most notably, the Mellor–Yamada turbulence closure scheme.[20] POM has been previously used in wide applications from estuaries[21] to coastal ocean circulation[22,23] to operational ocean forecasting.[24,25] In order to include the effect of waves on currents, the POM has been modified in two aspects: (1) the inclusion of a radiation stress;[2,6,26] and (2) the inclusion of a wetting and drying scheme.[27,28] Oey[29] developed a wetting and drying scheme independently. The wetting and drying grid scheme enables the POM to simulate the inundation process accompanying coastal storms, especially tropical storms. Applications of the modified POM to simulate storm surge and inundation in and around the Croatan–Albemarle–Pamlico Estuary system of North Carolina are described in Peng et al.[26] and in the Charleston Harbor, South Carolina in Peng et al.[30]

2.4. Parametrization of air-sea fluxes

2.4.1. The wave state and sea spray affected sea surface roughness

In atmospheric models, air-sea momentum flux is usually estimated through the Charnock relation[31]

$$gz_0/u_*^2 = \alpha, \tag{2}$$

where g is gravity, z_0 is the sea surface aerodynamic roughness, u_* is the friction velocity, and α is the Charnock constant which can also be thought of as the nondimensional roughness. In WRF the Charnock constant was chosen as 0.0185 following Wu,[32] which was widely used in other atmospheric and wave models. However, the classical Charnock relation does not explicitly consider wave state effects on sea surface roughness, though it has been commonly recognized that wave state has an important influence on air-sea momentum flux.[7,33–35] Also, recent studies[9,36–38] find that the classical Charnock relation is not applicable to high wind conditions when sea spray effects are significant. In order to consider both wave state and sea spray effects on sea surface wind stress, by combining the SCOR (Scientific Committee on Oceanic Research) relation[39] with the resistance law of Makin,[38] a parametrization of sea surface aerodynamic roughness applicable to both low-to-moderate and high winds is obtained:[40]

$$\frac{gz_0}{u_*^2} = \begin{cases} (0.085\beta_*^{3/2})^{1-1/\omega}[0.03\beta_* \exp(-0.14\beta_*)]^{1/\omega}, & \sim 0.35 < \beta_* < 35 \\ 17.60^{1-1/\omega}(0.008)^{1/\omega}, & \beta_* \geq 35 \end{cases}, \tag{3}$$

where $\beta_* = c_p/u_*$ is the wave age, and $\omega = \min(1, a_{cr}/\kappa u_*)$ is the correction parameter indicating the influence of sea spray on the logarithmic wind profile with κ the Karman constant, and the critical value of terminal fall velocity of the droplets $a_{cr} = 0.64\,\mathrm{m\,s^{-1}}$.[38] Equation (3) is used to parametrize air-sea momentum flux in the coupled system in which both wave state and sea spray effects are included. In addition, the roughness under smooth surface due to molecular viscosity $z_s = 0.11\nu/u_*$, where ν is the kinematic molecular viscosity of air depending on air temperature, is added to the sea surface roughness.[41]

As for the sea surface heat and moisture fluxes, the parametrization of sea surface scalar roughness parameters from COARE (Coupled Ocean Atmosphere Response Experiment) algorithm V3.1:[42]

$$z_T = z_q = \min(1.1 \times 10^{-4}, \quad 5.5 \times 10^{-5}\mathrm{Re}_*^{-0.6}), \tag{4}$$

where $\mathrm{Re}_* = z_0 u_*/\nu$ is the Reynolds number of sea surface aerodynamic roughness, is used to estimate the direct air-sea sensible and latent heat fluxes.

2.4.2. *The dissipative heating*

The frictional dissipation of atmospheric kinetic energy finally occurs at molecular scales, which in turn is converted into thermal energy. Following Zhang and Altshuler[43] the dissipative heating in the lowest level of the atmospheric model is expressed as

$$\left.\frac{dT}{dt}\right|_{Dis} = \frac{V_a u_*^2}{C_p z_1}, \tag{5}$$

where C_p is the air specific heat at constant pressure, z_1 is the height of model surface layer, and V_a is the wind speed at the model lowest semi-sigma layer. From Eq. (5) one can see that the dissipative heating is approximately proportional to the cubic power of surface wind speed. Thus under high winds, especially typhoon or hurricane conditions, dissipative heating increases rapidly with wind speed, which in turn will strengthen the typhoon or hurricane system. In the present coupled system, as we only consider the dissipative heating in the atmospheric surface layer,[43,44] it is equivalent to consider an upward sensible heat flux $H_E = \rho C_p V_a u_*^2$ at the bottom of the surface layer.

2.4.3. The wave state affected sea spray heat flux

Another important issue related to surface heat flux under high winds is the sea spray heat flux. To estimate the sea spray heat flux, one needs to know the sea spray generation function (SSGF) dF/dr_0, which quantifies how many spray droplets of initial radius r_0 are produced per square meter of the surface per second per micrometer increment in droplet radius. As to the SSGF for bubble-derived droplets, we introduce the whitecap coverage function of Zhao and Toba[45] into Monahan et al.'s SSGF,[46] thus obtaining a windsea Reynolds number (R_B) dependent SSGF:

$$\frac{dF}{dr_0} = 0.506 R_B^{1.09} r_0^{-2.95} (1 + 0.029 r_0^{1.02}) \times 10^{1.19 \exp(-B_0^2)}$$

$$B_0 = (0.666 - 0.976 \log r_0)/0.650 \tag{6}$$

Combining this SSGF applicable to bubble-derived droplets with Zhao et al.'s SSGF for spume droplets:[47]

$$\frac{dF}{dr_0} = 7.84 \times 10^{-3} R_B^{1.5} r_0^{-1}, \quad 30 < r_0 < 75\,\mu m$$

$$\frac{dF}{dr_0} = 4.41 \times 10^1 R_B^{1.5} r_0^{-3}, \quad 75 < r_0 < 200\,\mu m \,, \tag{7}$$

$$\frac{dF}{dr_0} = 1.41 \times 10^{13} R_B^{1.5} r_0^{-8}, \quad 200 < r_0 < 500\,\mu m$$

by filling the droplet radius gap between 20 and $30\,\mu m$ through interpolation, one can thus obtain a wave state affected SSGF applicable to both bubble-derived droplet and spume droplet.

Concerning the sea spray droplet microphysics,[48,49] using Andreas's method[50] to estimate the "nominal" sea spray sensible and latent heat fluxes, and further considering the following feedback effects:

$$H_{S,T} = H_S + \beta Q_S - \alpha \gamma Q_L$$

$$H_{L,T} = H_L + \alpha Q_L \tag{8}$$

where α, β, and γ are non-negative feedback coefficients, we thus can estimate the net sea spray contribution to the total sensible and latent heat fluxes

$$Q_{S,sp} = \beta Q_S - \alpha \gamma Q_L$$

$$Q_{L,sp} = \alpha Q_L \tag{9}$$

which are called sea spray sensible and latent heat fluxes hereinafter. α and γ are determined following Bao *et al.*[51] while β is taken as 1.[50] This method is used in the coupled modeling system to investigate the impacts of sea spray heat flux on hurricane intensity.

2.5. *Wave–current coupling*

Three types of wave effects are incorporated into the POM: (1) wave-induced surface wind stress; (2) wave-induced bottom stress; and (3) radiation stress. Contemporaneously, current fields (depth-averaged current) and surface wave level computed by the POM are provided to the wave model (SWAN or WW3). In other words, the coupling procedure in the present study is two-way and dynamic. So, values of currents and water elevation computed from POM are used in SWAN or WW3 to compute the wave parameters. Then the wave parameters are used to compute new surface wind stress, bottom stress and radiation stress, which are, in turn used in POM to compute the currents and surface water level at the next time step, and so on. Through this process, the two models are dynamically coupled. In this two-way coupling framework, POM and SWAN (or WW3) run using different time step and exchange data between them every 15 minutes, a common multiple of the two time steps. More details of the wave–current coupling procedure are described in Xie *et al.*[6]

2.6. *The coupling between WRF and POM*

In the outermost domain, the resolution of the POM (Fig. 2) is identical to that of the WRF model, thus the wind forcing is directly derived from the coupled atmosphere-wave simulations without interpolation. For the middle and the inner domains, surface stresses are interpolated from the outer domain grids through a linear interpolation scheme. In the current simulation, POM and WRF are only coupled through atmosphere–wave coupling. The wind stress computed by the coupled atmosphere–wave modeling system is provided to the POM.

It should be noted that in the present study, the POM is used to simulate the storm surge. The sea surface temperature and salinity fields are not simulated by the storm surge version of the POM. Instead, the WRF model uses the analyzed SST provided by the GFS data for computing air-sea fluxes, thus bypassing the POM for oceanic thermodynamic input parameters.

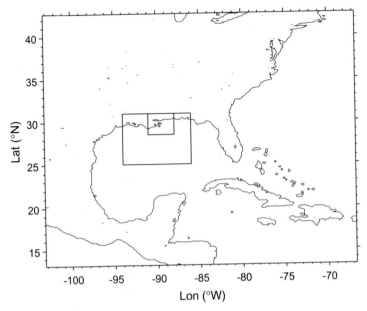

Fig. 2. Model domain and triple nesting windows for the wave-current interaction experiments.

3. Application of the Coupled System to Hurricane Katrina

3.1. *Atmosphere-wave interaction*

3.1.1. *Experiment design*

Hurricane Katrina is simulated by using the coupled atmosphere–wave modeling system to investigate the effects of air-sea coupling on hurricane system. The simulated period is from 0000 UTC 27 August to 0000 UTC 30 August, 2005. The grid spacing is 12 km. The coupled system exchanges variables between WRF and WW3 every 15 minutes, and the modeling results are output at 3 h interval.

The WRF model contains 298 × 275 horizontal grid points, and 30 sigma layers in the vertical direction. The integrating time step is 60 s. The WSM5 microphysics scheme,[52] Kain–Fritsch cumulus scheme,[53,54] YSU PBL scheme, and Dudhia short wave[55] and RRTM long wave[56] radiation scheme are chosen in this case study. The initial and lateral boundary conditions and sea surface temperature are provided by NCEP GFS 1° × 1° analysis data. The WW3 model resolves 32 frequencies logarithmically spaced from 0.0418 to 0.8023 Hz and 24 direction bands of 15 degrees each.

The wave model grid corresponds to the mass grid of the WRF model, and the time step is 15 minutes.

Two experiments are conducted to investigate the effects of coupling between atmosphere and surface waves on hurricane system. The control run (CTRL), with WRF and WW3 being uncoupled, utilize the classical Charnock relation to parametrize the air-sea momentum flux, and does not consider dissipative heating and sea spray heat flux. Whereas, the fully coupled run (CPL) couples WRF to WW3 through the wave state and sea spray affected surface roughness parametrization. The atmospheric surface layer dissipative heating and wave state affected sea spray heat flux are also included in the CPL experiment. The summary of experiments CTRL and CPL is shown in Table 1.

3.1.2. Results

The simulated sea level pressure, surface wind, and significant wave height (SWH) from the control experiment (CTRL) and the coupled (CPL) experiment are shown in Figs. 3(a)–3(f). In general the results from the CTRL and the CPL experiments look quite similar. There is no noticeable difference in storm location or spatial pattern of wind and pressure. However, the minimum center pressure of the storm simulated by the CPL experiment is lower than that of the CTRL experiment. Correspondingly, the maximum wind speed in the CPL experiment is stronger than that in the CTRL experiment. Since the simulated peak wind speed in both experiments are lower than that of the observed value, the stronger winds simulated by the coupled model suggests an improvement.

It is interesting to note that the spatial extend of the large wind area (e.g. wind speed larger than $25 \, \mathrm{m \, s^{-1}}$) is broader in the CTRL experiment than in the CPL experiment (Figs. 3(c) and 3(d)). As a result, the peak significant wave height in the CPL experiment and that in the CTRL experiment show comparable value, despite stronger maximum wind speed in the CPL experiment. These characteristics are also evident in the time series plots of minimum sea level pressure (SLP) (a), maximum

Table 1. Summary of the control (CTRL) experiment and the coupled (CPL) experiment.

Expts.	Aerodynamic roughness	Dissipative heating	Sea spray heat flux
CTRL	Equation (2)	No	No
CPL	Equation (3)	Yes	Yes

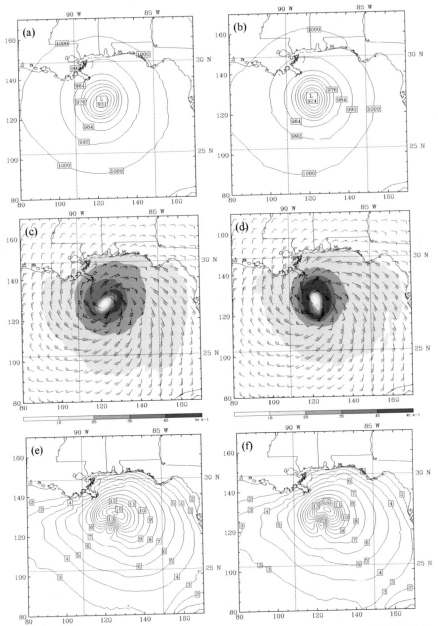

Fig. 3. Simulation results at 51 h. (a) sea level pressure (contour interval of 8 hPa) for the CTRL; (b) same as (a) but for the CPL; (c) 10 m wind vector (one full wind barb = $5\,\mathrm{m\,s^{-1}}$) and 10 m wind speed (contour interval of $10\,\mathrm{m\,s^{-1}}$) for the CTRL; (d) same as (c) but for the CPL; (e) significant wave height (contour interval of 1 m) for CTRL; (f) same as (e) but for CPL.

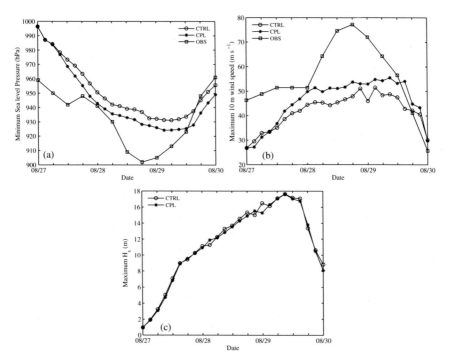

Fig. 4. Time series of (a) minimum sea level pressure; (b) maximum 10 m wind speed; and (c) maximum significant wave height for experiments CTRL (circle), CPL (asterisk), and observation (square).

Table 2. The simulated minimum sea level pressure, maximum 10-m wind, and maximum significant wave height (SWH) for experiments CTRL and CPL, together with the observational minimum central pressure and maximum surface wind speed.

Expts.	Min SLP (hPa)	Max 10-m wind ($m s^{-1}$)	Max SWH (m)
CTRL	931.4	48.37	17.2
CPL	924.3	54.36	17.0
OBS	902.0	77.17	—

10 m wind speed (b), and maximum significant wave height (c), as shown in Fig. 4.

Table 2 lists the simulated and observed minimum SLP and maximum 10 m 1 minute sustained wind, as well as the simulated maximum SWH. It should be noted that there are intensity phase errors for both CTRL and CPL (Figs. 3(a) and 3(b)), which is consistent to other NWP models' simulation. The observed peak intensity is around 1800 UTC 28 August, while the simulated max intensity is around 0300 UTC 29 August (51 h). The simulated minimum SLP from the CPL run (924 hPa) is approximately 7 hPa lower than that from the CTRL run (931 hPa). The maximum 10 m wind speed from the CPL run (54 m s^{-1}) is approximately 6 m s^{-1} higher than that from the control run (48 m s^{-1}). Although the maximum 10 m wind speed of CPL run is higher than that of CTRL run, the area with high wind speed (e.g. larger than 25 m s^{-1}) for the CPL run is smaller than that for the CTRL run as shown in Fig. 3. As a result, the simulated SWH from the CPL run is roughly the same as in the CTRL run. It should be noted that the coupled model is configured for a relatively low horizontal resolution (12 km grid), and the initialization is directly derived from the $1° \times 1°$ GFS analysis without using a bogusing vortex, the simulated hurricane is weaker than the observed storm in both the CTRL and the CPL experiments. In future studies and real-time forecasting, it is necessary to use a higher resolution for the WRF model in order to simulate the correct peak intensity of tropical cyclones. In the present study, since the goal is to test the model components and the coupling procedure, it is suffice to use the relatively low resolution of 12 km grid.

Figure 5 gives the simulated 51 h friction velocity (a), sea surface aerodynamic roughness (b), the relation between the drag coefficient and 10 m wind speed (c), and the relationship between the Charnock parameter and wave age β_* for the CTRL experiment. The corresponding fields from the CPL experiment are shown in Fig. 6. Comparison between the two experiments shows that the wave state and sea-spray induced sea surface roughness effect led to an increase of surface roughness due to wave age effect under low to moderate winds, but a decrease of surface roughness due to sea spray effect under high winds ($U_{10} > 25$ m s^{-1}). The maximum sea surface roughness is located at the areas where the wind speed is in the range of 25–35 m s^{-1} and the wave age is relatively small. Figure 6(c) shows that the drag coefficient is no longer linearly dependent upon wind speed because of the wave age effects. The impact of sea spray reduces the drag coefficient under high wind conditions. Figure 6(d) illustrates the simulated relation between the Charnock parameter and wave age, where the dots correspond to wind speed less than 25 m s^{-1} and circles correspond to wind speed larger than 25 m s^{-1}. It is evident that sea surface roughness decreases with wave age when wave age β_* is larger than about 15, and the existence of sea spray under high winds significantly reduces the sea surface roughness.

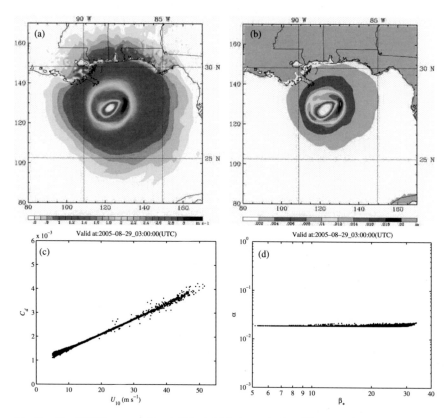

Fig. 5. The CTRL run results at 51 h. (a) friction velocity; (b) sea surface aerodynamic roughness; (c) relation between the drag coefficient and 10 m wind speed; and (d) relation between the Charnock parameter and wave age (β_*).

Figure 7 shows the simulated heat fluxes at 51 h, with (a) showing the total sensible heat flux $H_{S,T}$ from the CPL experiment; (b) the total latent heat flux $H_{L,T}$ from the CPL experiment; (c) the direct sensible heat flux H_S from the CPL experiment; (d) the direct latent heat flux H_L from the CPL experiment; (e) the same as (c) but for the CTRL experiment; (f) the same as (d) but for the CTRL experiment. Overall, the sensible and latent heat fluxes show large values in the region of maximum wind speed in both CPL and CTRL experiments. The effect of sea spray on air-sea sensible and latent heat fluxes is shown in Figs. 8(a) and 8(b). From Fig. 8(a), one can see that the sea spray sensible and latent heat fluxes are significant in areas with strong winds and waves. The sea spray induced sensible heat flux is negative (Fig. 8(a)), thus making a negative contribution to the

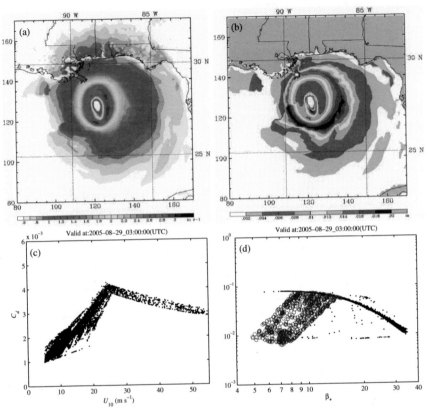

Fig. 6.   Same as Fig. 5 but for CPL. The blue dots in (d) correspond to 10 m wind
speed less than $25\,\mathrm{m\,s^{-1}}$, while the red circles correspond to 10 m wind speed larger than
$25\,\mathrm{m\,s^{-1}}$.

total upward sensible heat flux (Fig. 7(a)). Whereas, the sea spray induced
latent heat flux (Fig. 8(b)) shows a positive contribution to the total latent
heat flux (Fig. 7(b)). Comparing experiment CTRL with CPL, sea spray
induced heat flux increases the direct sensible heat flux, but reduces the
direct latent heat flux. This is because evaporation of the spray droplets
reduces low-level air temperature and increases low-level moisture, thus
increases air-sea temperature difference and reduces the air-sea moisture
difference. Figure 8(c) shows the CPL simulated 51 h equivalent sensible
heat flux to dissipative heating. The dissipative heating makes significant
contribution to the total air-sea heat flux in high wind areas, while in low
wind areas its contribution is negligible.

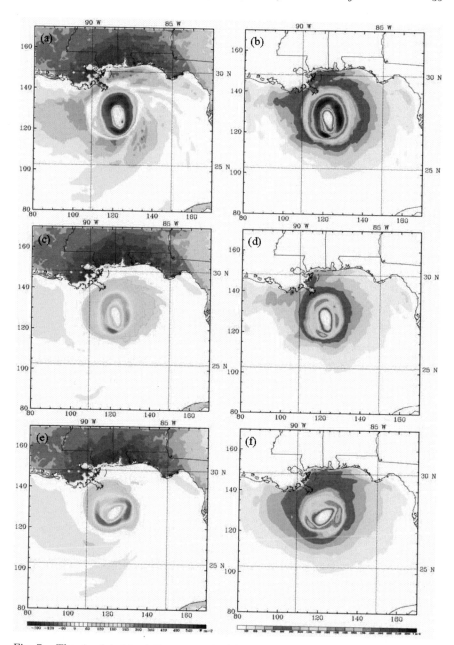

Fig. 7. The simulated heat fluxes at 51 h. (a) total sensible heat flux $H_{S,T}$, (b) total latent heat flux $H_{L,T}$, (c) direct sensible heat flux H_S, and (d) direct latent heat flux H_L for the CPL experiment; (e) same as (c) but for the CTRL experiment; (f) same (d) but for the CTRL experiment.

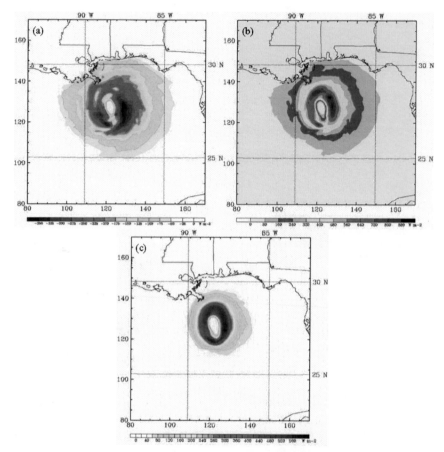

Fig. 8.   The simulated effect of sea spray on air-sea fluxes in the CPL experiment at 51 h. (a) sea spray sensible heat flux; (b) sea spray latent heat flux; and (c) dissipative heating (H_R).

3.2. Storm surge

3.2.1. Model configuration

In this case study, the wave–current coupled system is configured for three nested domains as shown in Fig. 2 with the smallest imbedded into a middle-sized domain, which, in turn, imbedded into a larger one. The outermost grid is the same as the wave model grid corresponding to the mass grids of the WRF model with a 12 km spatial grid size. The middle domain spans 94.0–86.0°W, 25.0–31.0°N, with a 3.6 km spatial grid size. The innermost domain covers 91.0–88.0°W, 28.5–31.0°N, with a 916 m spatial grid size.

Four uniformly spaced sigma levels are used in the vertical for all domains. A minimum depth of 1 m was given to grid cells where the mean water depth is less than 1 m.

The POM is driven by surface wind stress provided by the coupled wind-wave model discussed in Sec. 3.1. Through wave–current coupling, the POM also computes the bottom stress and the radiation stress from the wave parameters provided by the wave model.

3.2.2. *Results*

Four locations are chosen for time series data comparisons (Fig. 9). As shown in Fig. 10, except at location 4, the coupling between waves and surge produces a 10–40 cm increase in peak storm surge, which represents a 5–20% modification to the peak positive surge. The effect of waves on the peak negative storm surge (Figs. 11(b) and 11(c)) appears to be larger than for the peak positive surge (Figs. 11(a) and 11(d)). Ocean surface waves significantly increased the absolute values of the peak negative surge at location 2 and 3 at around 60 hours. Spatially, the wave-induced increases of surge occurred mainly in regions influenced by onshore wind stress, whereas increases in negative surge occurred in regions where waves caused

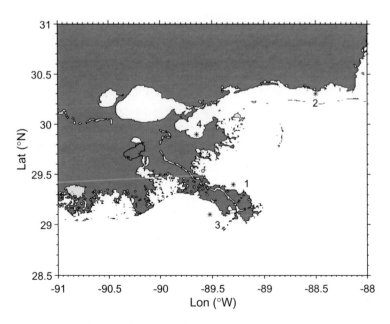

Fig. 9.   Locations of observing stations 1–4.

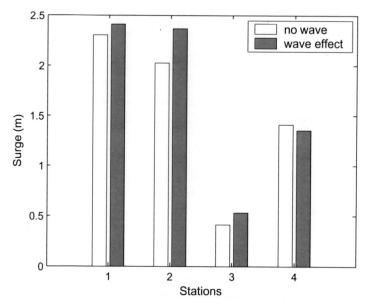

Fig. 10.   Comparison of simulation results at the 4 observing stations (Fig. 9) for the coupled and the uncoupled simulations. Open bars are for uncoupled run, solid bars are for the coupled run.

an increase in off-shore wind stress (Fig. 12). It should be noted that the simulated peak surge is generally less than 3 m, which is less than the 4–5 m peak surge (http://www.nhc.noaa.gov) reported for Hurricane Katrina. This is clearly the result of the weaker than observed wind speed simulated by the WRF model, as shown in Figs. 3 and 4. In order to simulate the correct magnitude of the peak surge, it is necessary to for the atmospheric model to accurately simulate the intensity of the storm. At the current 12 km grid size, the coupled modeling system under predicts the magnitude of the wind speed and the storm surge.

4. Summary and Discussions

A regional modeling system is constructed to simulate wind–wave–current interactions under hurricane conditions. The coupled system consists of the Weather Research and Forecasting (WRF) regional model; a regional ocean circulation model (POM); and a third-generation ocean surface wave model (SWAN or WW3). These component models are coupled through two-way property exchanges via the Model Coupling Toolkit (MCT).

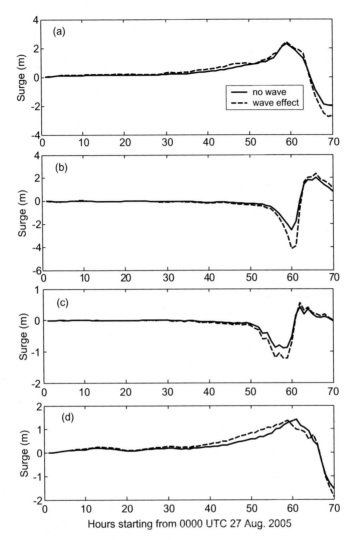

Fig. 11.   Comparison of storm surge time series between the coupled and the uncoupled simulations at the 4 locations shown in Fig. 9. (a) station 1; (b) station 2; (c) station 3; (d) station 4. Solid curve: uncoupled surge run; dash curve: coupled run.

The modeling system has been tested for both stand-alone and coupled simulations of Hurricane Katrina, as well as oceanic responses of surface waves and storm surge. Coupled simulations reveal significant contributions from air-sea and wind-wave coupling, as well as sea spray effects to the intensity change of the hurricane, storm surge, and waves induced by

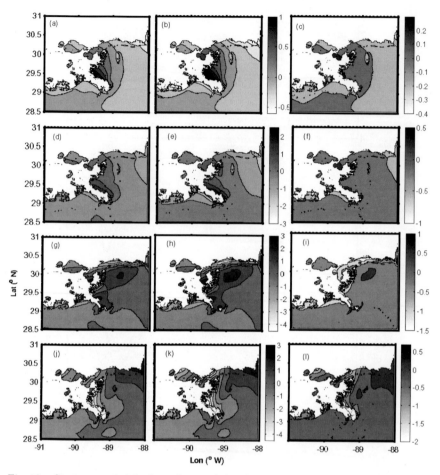

Fig. 12.   Storm surge height from the coupled and the uncoupled runs. (a), (d), (g) and
(j): coupled run at 0100 UTC, 0700 UTC, 1300 UTC and 1900 UTC, respectively, on
29 August 2005; (b), (e), (h) and (k): same as (a), (d), (g), (j), but for the uncoupled
run; and (c), (f), (i) and (l): the difference of surge height between the coupled and the
uncoupled run.

the storm. However, the test case was run on a relatively coarse grid of
12 km for the atmosphere initialized with $1° \times 1°$ GFS analysis. As a
result, the peak storm intensity of the simulated hurricane is lower than
the observed peak intensity. Thus, although the coupled modeling system
showed a significant improvement in the simulation of the hurricane, the
simulated wind speed is lower than the observed wind speed, which, in turn,
led to lower-than-observed peak storm surge and significant wave height.

Nevertheless, taking the integrated approach to simulating and ultimately forecasting the wind–wave–current system is a step in the right direction toward a dynamically consistent simulation of the coupled wind–wave–current system that is present in the real world.

Acknowledgments

This study is a joint effort between North Carolina State University and the Ocean University of China. The project is partially supported by grants awarded by the U.S. National Oceanic and Atmospheric Administration (NOAA) through subcontract #UF-EIES-0704029NCS and U.S. Department of Energy award #DE-FG02-07ER64448. The lead author also acknowledges the support of a Changjiang Distinguished Professorship awarded by China Ministry of Education.

References

1. L. Xie, in *Observation, Theory and Modeling of Atmospheric Variability*, ed. X. Zhu (World Scientific Publishing, Singapore, 2004), p. 612.
2. L. Xie, K. Wu, L. J. Pietrafesa and C. Zhang, *J. Geophys. Res.* **106** (2001) 16,841.
3. G. L. Mellor, *J. Phys. Oceanogr.* **33** (2003) 1978.
4. G. L. Mellor, *J. Phys. Oceanogr.* **35** (2005) 2291.
5. H. Xia, Z. Xia and L. Zhu, *Coastal Eng.* **51** (2004) 309.
6. L. Xie, H. Liu and M. Peng, *Ocean Model.* **20** (2008) 252.
7. M. A. Donelan, in *The Sea: Ocean Engineering Science*, eds. B. l. Mehaute and D. M. Hanes (Wiley-Interscience, 1990), pp. 239–292.
8. C. Guan and L. Xie, *J. Phys. Oceanogr.* **34** (2004) 2847.
9. M. D. Powell, P. J. Vickery and T. A. Reinhold, *Nature* **422** (2003) 279.
10. W. C. Skamarock, J. B. Klemp, J. Dudhia, D. O. Gill, D. M. Barker, W. Wang and J. G. Powers, A description of the advanced research WRF, version 2, NCAR Technical Note NCAR/TN-468+STR (2005).
11. N. Booij, R. C. Ris and L. H. Holthuijsen, *J. Geophys. Res.* **104** (1999) 7649.
12. H. L. Tolman, User manual and system documentation of WAVEWATCH-III version 2.22, NOAA/NWS/NCEP/MMAB Technical Note 222(2002), p. 133.
13. J. Larson, R. Jacob and E. Ong, *Int. J. High Performance Comput. Appl.* **19** (2005) 277.
14. R. Jacob, J. Larson and E. Ong, *Int. J. High Performance Comput. Appl.* **19** (2005) 293.
15. A. Craig, R. Jacob, B. Kauffman, T. Bettge, J. Larson, E. Ong, C. Ding and Y. He, *Int. J. High Performance Comput. Appl.* **19** (2005) 309.
16. J. C. Warner, N. Perlin and E. D. Skyllingstad, *Environ. Model. Software* **23** (2008) 1240.

17. W. Wang, D. Barker, J. Bray, C. Bruyère, M. Duda, J. Dudhia, D. Gill and J. Michalakes (2007), http://www.mmm.ucar.edu/wrf/users/docs/user_guide/.
18. G. J. Komen, L. Cavaleri, M. Donelan, K. Hasselmann, S. Hasselmann and P. A. E. M. Janssen, *Dynamics and Modelling of Ocean Waves* (Cambridge University Press, Cambridge, 1994), p. 532.
19. G. L. Mellor and A. F. Blumberg, *Mon. Weather Rev.* **113** (1985) 1380.
20. G. L. Mellor and T. Yamada, *Rev. Geophys.* **20** (1982) 851.
21. L.-Y. Oey, G. L. Mellor and R. I. Hires, *J. Phys. Oceanogr.* **15** (1985) 1676.
22. A. F. Blumberg and G. L. Mellor, *J. Geophys. Res.* **88** (1983) 4579.
23. A. F. Blumberg and G. L. Mellor, in *Three-Dimensional Coastal Ocean Models*, ed. N. Heaps (American Geophysical Union, 1987), p. 208.
24. F. Aikman, G. L. Mellor, T. Ezer, D. Shenin, P. Chen, L. Breaker and D. B. Rao, in *Modern Approaches to Data Assimilation in Ocean Modeling*, ed. P. Malanotte-Rizzoli (1996), pp. 347–376.
25. F. Aikman, III and D. B. Rao, in *Coastal Ocean Prediction, Coastal and Estuarine Studies, Vol. 56*, ed. C. N. K. Mooers (American Geophysical Union, Washington, 1999), pp. 467–499.
26. L. Xie, L. J. Pietrafesa and K. Wu, *J. Geophys. Res.* **108** (2003) 3049.
27. L. Xie, L. J. Pietrafesa and M. Peng, *J. Coastal Res.* **20** (2004) 1209.
28. M. Peng, L. Xie and L. J. Pietrafesa, *Estuarine, Coastal and Shelf Science* **59** (2004) 121.
29. L.-Y. Oey, *Ocean Model.* **9** (2005) 133.
30. M. Peng, L. Xie and L. J. Pietrafesa, *J. Geophys. Res.*, vol. 111 (2006), doi:10.1029/2004JC002755.
31. H. Charnock, *Quart. J. Roy. Meteor. Soc.* **81** (1955) 639.
32. J. Wu, *J. Phys. Oceanogr.* **13** (1980) 1441.
33. Y. Toba, N. Iida, H. Kawamura, N. Ebuchi and I. S. F. Jones, *J. Phys. Oceanogr.* **20** (1990) 705.
34. H. K. Johnson, J. Hojstrup, H. J. Vested and S. E. Larsen, *J. Phys. Oceanogr.* **28** (1998) 1702.
35. W. M. Drennan, H. C. Graber, D. Hauser and C. Quentin, *J. Geophys. Res.* **108** (2003) 8062, doi:10.1029/2000JC000715.
36. M. Alamaro, K. A. Emanuel, J. J. Colton, W. R. McGillis and J. Edson, *25th Conf. Hurricanes and Tropical Meteorology*, San Diego, California (2002), p. 2.
37. M. A. Donelan, B. K. Haus, N. Reul, W. J. Plant, M. Stiassnie, H. C. Graber, O. B. Brown and E. S. Saltzman, *Geophys. Res. Lett.*, vol. 31 (2004), doi:10.1029/2004GL019460.
38. V. K. Makin, *Boundary-Layer Meteorol.* **115** (2005) 169.
39. I. S. F. Jones and Y. Toba, *Wind Stress over the Ocean* (Cambridge University Press, 2001), p. 307.
40. B. Liu, C. Guan and L. Xie, *28th Conf. Hurricanes and Tropical Meteorology*, Orlando, Florida (2008).
41. S. D. Smith, *J. Geophys. Res.* **93** (1988) 15467.
42. C. W. Fairall, E. F. Bradley, J. E. Hare, A. A. Grachev and J. B. Edson, *J. Climate* **16** (2003) 571.
43. D.-L. Zhang and E. Altshuler, *Mon. Weather Rev.* **127** (1999) 3032.

44. M. Bister and K. A. Emanuel, *Meteor. Atmos. Phys.* **65** (1998) 233.
45. D. Zhao and Y. Toba, *J. Oceanogr.* **57** (2001) 603.
46. E. C. Monahan, in *The Role of Air-Sea Exchange in Geochemical Cycling*, eds. P. Buat-Menard and D. Reidel (Dordrecht, 1986), pp. 129–163.
47. D. Zhao, Y. Toba, K.-I. Sugioka and S. Komori, *J. Geophys. Res.*, vol. 111 (2006), doi:10.1029/2005JC002960.
48. E. L. Andreas, Thermal and size evolution of sea spray droplets, CRREL Report 89–11 (1989), p. 37.
49. E. L. Andreas, *Tellus, Ser. B.* **42** (1990) 481.
50. E. L. Andreas, *J. Geophys. Res.* **97** (1992) 11429.
51. J.-W. Bao, J. M. Wilczak, J.-K. Choi and L. H. Kantha, *Mon. Weather Rev.* **128** (2000) 2190.
52. S.-Y. Hong, J. Dudhia and S.-H. Chen, *Mon. Weather Rev.* **132** (2004) 103.
53. J. S. Kain and J. M. Fritsch, *J. Atmos. Sci.* **47** (1990) 2784.
54. J. S. Kain and J. M. Fritsch, in *The Representation of Cumulus Convection in Numerical Models*, eds. K. A. Emanuel and D. J. Raymond (American Meteorology Society, 1993), p. 246.
55. J. Dudhia, *J. Atmos. Sci.* **46** (1989) 3077.
56. E. J. Mlawer, S. J. Taubman, P. D. Brown, M. J. Iacono and S. A. Clough, *J. Geophys. Res.* **102** (1997) 16663.

Advances in Geosciences
Vol. 18: Ocean Science (2008)
Eds. Jianping Gan et al.
© World Scientific Publishing Company

COUPLED DATA ASSIMILATION
FOR ENSO PREDICTION*

DAKE CHEN[1,2]

[1] *State Key Laboratory of Satellite Ocean Environment Dynamics, Hangzhou, China*
[2] *Lamont-Doherty Earth Observatory of Columbia University, New York, USA*
dchen@sio.org.cn

An outstanding problem with present ENSO forecast systems is that most of them are initialized in an uncoupled manner, that is, no feedbacks are allowed between the ocean and the atmosphere during data assimilation and model initialization. Such an approach may produce realistic initial states, but not necessarily the optimal conditions for skillful forecasts, because model-data mismatch can cause serious initialization shock. A more reasonable approach is to initialize forecast systems using a coupled approach, which assimilates data, both oceanic and atmospheric, into the coupled models that are used for forecast. Here we briefly review the progress in this important research area. In particular, based on the evolution history of an intermediate couple model, and some preliminary results from a state-of-the-art forecast system, we demonstrate the impact and necessity of coupled data assimilation, and we suggest it may hold a key for further improvement of the predictive skill of present ENSO models.

1. Introduction

The last few decades have seen tremendous advances in climate research and prediction. Among all identified climate modes in the Earth's climate system, El Niño–Southern Oscillation (ENSO) has so far shown the highest predictability and, because of its far-reaching influences, predictions of ENSO-related tropical SST anomalies have become the basis for global seasonal forecasts of surface temperature and precipitation. It is largely due to the measurable predictability of ENSO and the quantification of its global impact that climate prediction is no longer a speculative practice [1, 2].

---

*This paper is mostly based on an invited presentation at the 2009 annual meeting of Asia Oceania Geosciences Society (AOGS).

At present, a hierarchy of dynamical ocean-atmosphere coupled models, ranging from intermediate couple models (ICM) to coupled general circulation models (CGCM), is routinely used for ENSO forecasting (see a collection of them at http://iri.columbia.edu/climate/ENSO, the forecast website of IRI, the International Research Institute for Climate and Society). Most of these models have statistically significant skills in predicting indices of tropical Pacific sea surface temperature (SST) anomaly from several months to a year in advance. However, as shown by the El Niño Simulation Intercomparison Project [3], and by the recent Fourth Assessment Report (AR4) of the Intergovernmental Panel on Climate Change (IPCC), none of the models is able to realistically simulate all aspects of the interannual SST variability in the tropical Pacific, even if only basic diagnostics are considered. Despite their vast differences in complexity, many of the present ENSO models exhibit comparable predictive skills, which seem to have hit a plateau at moderate level.

Generally speaking, there are four factors that limit the current skill of ENSO prediction: inherent limits to predictability, gaps in observing systems, model flaws, and suboptimal use of available observational data [4]. There is still considerable debate on the inherent limits to predictability, but increasing evidence suggests that our current level of predictive skill is still far from those limits and surely there is plenty of room for improvement. Thus our main task is to improve observing systems, forecast models, and data assimilation methods. Enormous efforts have been made in all these areas in recent years. Observation networks such as the Tropical Atmosphere Ocean (TAO) buoy-mooring array and the altimeter and scatterometer satellite missions have proven invaluable for ENSO monitoring and forecasting [5, 6]; regional and global models of all ranks have been continuously improved in terms of both physics and computational capability [7, 8]; and various data assimilation schemes and model initialization procedures have been developed and applied to ENSO prediction. Our focus here is on this last area of research.

A major shortcoming of present ENSO forecast systems is that most of them are initialized in an uncoupled manner, that is, no feedbacks are allowed between the ocean and the atmosphere during data assimilation and model initialization. For both retrospective and operational forecasts, the initial conditions are often given by independently obtained oceanic and atmospheric reanalysis products. This approach may produce realistic initial states, but not necessarily the optimal conditions for skillful forecasts. The reason is that present models are still far from reality. Once a forecast

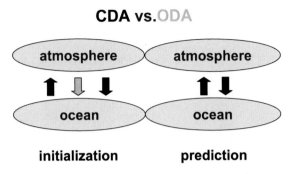

initialization   prediction

Fig. 1. Schematic showing the difference between CDA and ODA. During initialization, the grey arrow denotes the one-way communication via ODA, while the black arrows indicate the two-way ocean-atmosphere interaction via CDA. The latter is consistent with what happens during prediction.

starts, one has to rely on the biased model, and an "initialization shock" and swift "climate drift" would occur at the transition from uncoupled to coupled runs. A more natural approach is to initialize models using coupled data assimilation (CDA), that is, to assimilate data, both oceanic and atmospheric, into the coupled models that are used for forecast. The difference between CDA and the traditional ocean data assimilation (ODA) is illustrated in Fig. 1. Unfortunately, despite the obvious advantage of CDA, there have been only a few attempts on its implementation [9–12], and most of them are based on relatively simple ICMs.

The rest of this paper is organized as follows. In Secs. 2, 3 and 4, we discuss, respectively, the first successful use of CDA for ENSO prediction, the importance of model bias correction for effective CDA, and a unique application of CDA for the longest retrospective forecast ever made, all based on an ICM that has been widely used for ENSO research and prediction. Then in Sec. 5 we describe our current attempts to implement CDA to a state-of-the-art CGCM, followed by summary and conclusion in Sec. 6.

2. Lesson One: Wind Nudging

To our knowledge, the concept of CDA and its application to ENSO prediction was first proposed by Chen *et al.* [9]. The motivation of their study was to improve the predictive skill of the LDEO model, an ICM used at the Lamont-Doherty Earth Observatory (LDEO) of Columbia University, also known to the community as the Cane-and-Zebiak model [13, 14]. It is the first physics-based ocean-atmosphere coupled model designed for ENSO research, and has played an important historical role in our understanding and prediction of ENSO. By the mid-1990s, the model had been used to produce experimental forecasts on a monthly basis for a decade, but its forecast scheme remained the same and its skill was still far from satisfactory.

Although the simplified model physics could be partly responsible for the low predictive skill, the one-way initialization scheme used by the model at the time could be a more severe limitation. The initial conditions were obtained by first driving the oceanic component with observed surface wind and then using the resulting SST field to force the atmospheric component, without explicit consideration of the feedbacks between the ocean and the atmosphere. This could be problematic because of the mismatch between modeled and observed winds, which could cause an "initialization shock" when forecast starts, as indicated by the jumpy behavior and frequent false alarms of the original model. To overcome the problem, Chen *et al.* [9, 15] proposed a CDA procedure that assimilates wind data in a coupled manner and thus makes the initial conditions more balanced and self-consistent. Basically, instead of forcing the ocean model with observed winds, which is equivalent to inserting the full wind data into the model, this procedure modifies the surface wind field with a weighted average of modeled and observed winds. The premise is that the coupled model itself can produce realistic interannual variability of the ocean–atmosphere system if only slightly nudged toward observed winds.

Figure 2(a) compares the zonal wind stress anomalies at the equator from the cases with the standard (LDEO1) and new (LDEO2) initialization procedures. For both cases the wind stress shows large interannual oscillations, but the case with CDA is much less noisy. Figure 1(b) demonstrates the impact of this wind difference on the equatorial thermocline depth anomalies, which may be taken as a measure of the anomalous upper ocean heat content. Again, there is a striking difference between the two cases: the energetic high frequency fluctuations evident in the original procedure are largely eliminated with the new one. While the oceanic component generates high-frequency fluctuations when forced by the observed winds, the coupled model preferentially selects the low frequency, interannual variability. The CDA also results in a shallower thermocline in the western equatorial Pacific during most ENSO events, with implications for the termination of these warm episodes. The

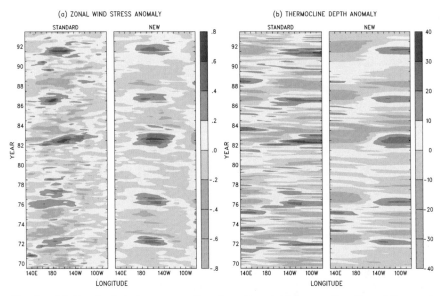

Fig. 2. (a) Zonal wind stress anomalies along the equator, as a function of time and longitude, for the standard and new cases. (Left) Observed wind stress anomalies used to initialize the standard model (LDEO1). (Right) Wind stress anomalies obtained using CDA for the new model (LDEO2). (b) Corresponding model thermocline depth anomalies in the two cases. Adopted from [9].

retrospective forecasts made with the two different sets of initial conditions are compared in Fig. 3 in terms of NINO3 index, the SST anomaly averaged over the region 5°S–5°N and 90°–150°W. The LDEO1 forecasts capture the onsets of large warming events more than one year in advance, but there are several false alarms in the intermediate periods and noticeable scatter, especially at longer lead times. The performance of LDEO2 is much improved, as the number and magnitude of erroneous forecasts have been reduced considerably in the periods between large warming events. In particular, the troublesome prolongation of the 1982–83 El Niño in LDEO1 is completely eliminated with the CDA initialization. A large fraction of the scatter among long-lead forecasts has also been eliminated.

By simply applying the CDA procedure without acquiring any more data than the wind stress used in the original scheme, the forecast skill of the LDEO model was improved over all lead times, with an increase in anomaly correlation of 0.1–0.3 and a reduction in root mean square error of 0.2–0.4°C. The seemingly inevitable "spring barrier" in predictability was also eliminated. The dramatic improvement came as a big surprise

FORECAST/OBSERVED NINO3 SST ANOMALY

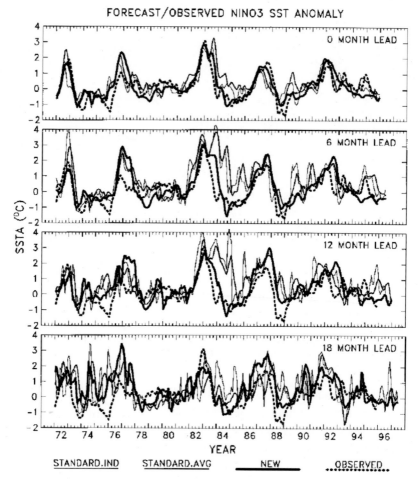

Fig. 3. Time series of observed and forecast NINO3 SST anomalies. Forecasts with 0-, 6-, 12-, and 18-month leads are shown in different panels for LDEO1 (standard) and LDEO2 (new) model forecasts. The observed anomalies are repeated from panel to panel. In each panel there are two curves for LDEO1: one for individual forecasts and the other for the averages of six consecutive forecasts. Only individual forecasts are shown for LDEO2. Adopted from [15].

to many and was considered a breakthrough in ENSO prediction. It is clear from Fig. 2 that CDA has the effect of filtering out high-frequency signals present in the original wind-forced initial conditions. This occurs because the dominant mode of variability in this model is ENSO-like, that is, large scale and low frequency. The high-frequency components of the initial conditions, which act as noise to the model, degrade forecast performance. By dynamically filtering out these components, the CDA procedure effectively reduces the mismatch between the observed initial conditions and the model's intrinsic variability, while retaining the large-scale, low-frequency information essential to ENSO. This simple CDA method was later extended to include sea level (LDEO3) and SST (LDEO5) data. It is possible that the same procedure may not work for other more complicated systems, but in principle the need for some form of CDA is undeniable, since neither models nor data will ever be perfect, and there will always be incompatibility between the two.

3. Lesson Two: Bias Correction

As discussed above, the purpose of CDA is to produce initial conditions that are consistent with the model's internal dynamical balance so that a smooth forecast start can be ensured. If the model has large systematic biases, however, this can only be achieved by assigning less weight to observational data, which is exactly what has been done in the wind nudging example given in the last section. Such an approach implicitly assumes that the model does not need much data to keep itself on track, evidently the case for the 1980s and early 1990s (Fig. 2), when the observed ENSO bore a strong resemblance to the model's internal variability. However, the assumption does not hold for recent years, when the model simply could not work well without a strong helping hand from data [16, 17]. In any case, a forecast initialization that neglects the majority of available observations cannot be optimal. Thus a prerequisite for an effective CDA is to eliminate or reduce systematic model biases.

Considering the tremendous effort on data assimilation, bias correction has not been given much attention in the past. The formalisms typically used in ocean and atmosphere data assimilation techniques take a "textbook" approach and assume that model biases do not exist. In practice, the "unbiased" *a priori* error estimates are often inflated in order to achieve consistency in a posteriori verification. Consequently, almost all successful uses of data assimilation in ENSO forecasting weight models unrealistically high compared to data. This is particularly true for adjoint methods, which treat the model as if it had zero error. Another way of describing the same problem is just what we mentioned earlier: there is a shock when data are inserted into initial model states without taking account of model biases. The adjoint methods take the ultimate path to remove it, sacrificing data if needed. In other schemes, the data-model difference projects onto rapidly growing error modes, resulting in a poor forecast.

Chen *et al.* [17] demonstrated that the systematic biases of the LDEO model can be effectively reduced with an interactive statistical correction based on the regression between the leading empirical orthogonal functions (EOFs) of the model errors and the leading multivariate EOFs (MEOFs) of the model states. As compared to the previous versions of the model, the bias-corrected model (LDEO4) not only exhibits a more realistic internal variability (Fig. 4), but also performs better in ENSO forecasting (Fig. 5). It is important to note that the bias correction is an integral part of the coupled model so that, for instance, the bias-corrected SST field will be

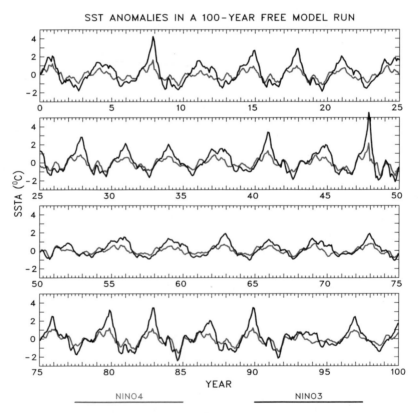

Fig. 4.   Time series of NINO3 and NINO4 SST anomalies from a 100-year free run of the bias-corrected LDEO4 model. The internal model variability bears strong resemblance to that observed.

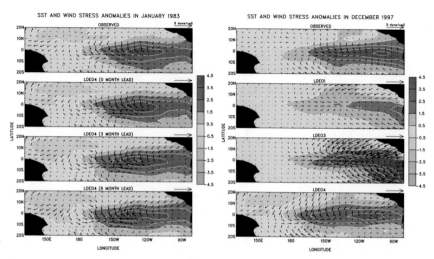

Fig. 5.   Left: Observed and forecast SST and wind stress anomalies in January 1983. Forecasts were made by LDEO4 at lead times of 0, 3, and 6 months. Right: Observed and forecast SST and wind stress anomalies in December 1997. Forecasts were made at 6 month lead by LDEO1, LDEO3 and LDEO4, respectively. The bias-corrected LDEO4 is clearly the winner. Adopted from [18].

used for the next computation of the wind field, and so on. Thus it is different from other commonly used corrections based on model output statistics (MOS) or interface flux climatology. A bias-corrected model can have a different and more realistic internal variability. With such a model, data assimilation is less sensitive to the choice of nudging parameters and is thus more straightforward. This kind of statistical bias correction procedure should be generally applicable to other coupled ocean-atmosphere models, albeit the specifications of the optimal statistical corrector may differ.

In recent years, the importance of bias correction in the context of CDA has been recognized. A workshop on CDA sponsored by the National Oceanic and Atmospheric Administration (NOAA) was held in the spring of 2003 to explore the possibility of implementing systematic data assimilation into CGCMs [12]. The workshop concluded that, for initialization of seasonal-to-interannual predictions, more research is needed into (1) best initialization compared with best analysis; (2) initializing coupled models; and (3) statistical correction to compensate for biases. CDA presents a host of problems quite different from those in data assimilation into the forced ocean component. Model biases are much harder to deal with in a coupled model than in a stand-alone component model and, if they are not properly corrected, initial errors would grow fast and largely degrade forecasts, which is particularly true for complex coupled systems. More research is definitely needed along this line, especially in the analysis of the pattern, nature, and statistics of the model biases, and in the implementation of proper and effective bias correction schemes into CGCMs.

4. Lesson Three: Century-Long Hindcasts

A major roadblock in ENSO prediction research is the lack of long enough hindcast experiments to assess model skill, to identify model deficiencies, and to study ENSO variability and predictability on various timescales. Most of the existing experiments of this sort only cover the last 10–30 years, with degrees of freedom too few to allow robust assessment of skill for interannual and longer-term fluctuations. Thus it is very desirable to extend such experiments all the way to the mid-19th century, when instrumental in-situ observations first became abundantly available. The main obstacle to this endeavor is the limitation of the historical data for adequate model initialization, and in part also the inability of present climate models to make effective use of available data. Presently, the only trustworthy oceanic datasets longer than a century are reconstructed SST products. The question is whether or not it is possible to carry out century-long hindcast experiments with only SST data assimilated. It is unlikely with ODA, but quite feasible with CDA, as shown by a series of recent studies [11, 19, 20].

The first long-term ENSO hindcast experiment with CDA for initialization was that of Chen et al. [11]. They performed an unprecedented set of hindcasts starting from every month in the past one and a half centuries, using the latest version of LDEO model (LDEO5) with only reconstructed SST data for model initialization. The model was able to predict most of the warm and cold events that occurred during this long period, especially the relatively large warm and cold events. A comparison of model predictions with observations is shown in Fig. 6 in terms of composites of all El Niños and La Niñas that have amplitudes greater than 1°C. Clearly the strength and spatial pattern of ENSO are well captured by the model. Figure 7 shows the long-lead retrospective forecasts of the 6 largest El Niño events since 1856. In all cases, the model predicted the observed strong El Niños two years in advance, though some errors exist in the forecasted onset and magnitude of these events. Since the LDEO model does not contain any internal high-frequency variability, its success

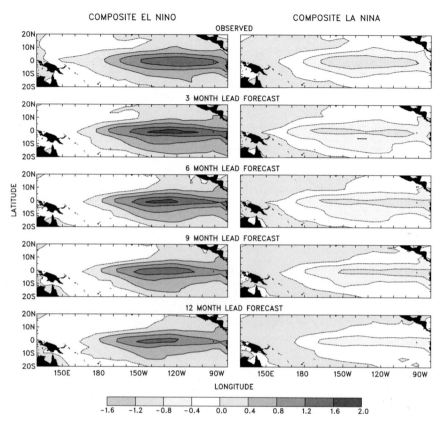

Fig. 6. Composite El Niño and La Niña from 24 warm events and 23 cold events over the 148-year period from 1856 to 2003. Top panels are observations [21] and the rest are predictions at different lead times by LDEO5 model. Adopted from [11].

implies that the evolution of major ENSO events is largely determined by oceanic initial conditions rather than unpredictable atmospheric "noise". These results favor the interpretation that the enhanced wind burst activity in the boreal spring preceding large El Niños is a consequence of those ongoing events [22] rather than a cause [23]. A practical consequence is a more optimistic notion for the possibility of skillful long-lead forecasts of El Niño.

It is also evident from this long-term hindcast experiment that there are considerable decadal and interdecadal variations in ENSO and its predictability [11]. These variations could not be due to differences in data coverage, as the most predictable period turned out to be 1876–1895,

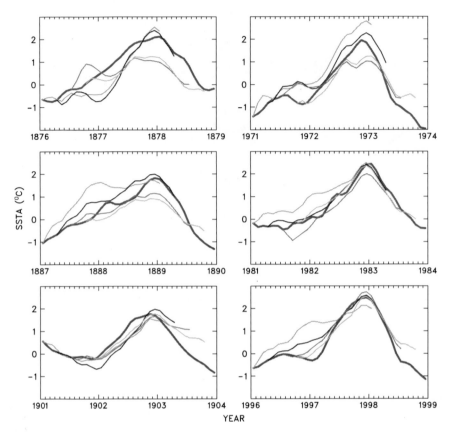

Fig. 7.   Six of the largest El Niños since 1856. The thick curves are observed NINO3.4
(5°S–5°N, 120°–170°W) SST anomalies and the think curves are predictions started 24,
21, 18 and 15 months before the peak of each El Niño, respectively. Adopted from [11].

when observational data were sparse. Tang *et al.* [19] further explored
the interdecadal variability using an ensemble of three ENSO forecast
models, including LDEO5 and two hybrid coupled models (HCM), for
the 120-year period of 1880–2000. The predictive skills of these models
showed a very consistent interdecadal variation, with high skill in the
late 19th century and in the mid-late 20th century, and low skill for
the period 1900–1960. The interdecadal variation in ENSO predictability
is in good agreement with that in the signal strength of interannual
variability and in the degree of ENSO asymmetry (which is a measure
of nonlinearity that makes the amplitude of El Niños greater than that
of La Niñas). High predictability is attained when the strength of ENSO

signal and its asymmetry are enhanced, and vice versa. Using a linear approximation of the 148-year hindcast experiment of LDEO5, Cheng *et al.* [20] investigated ENSO predictability with singular vector analysis, and identified the patterns of optimal initial error growth. All these studies were made possible by an effective CDA using historical SST data. Note that in the CDA initialization procedure used here, assimilating SST data is not simply putting a constraint on the ocean model with SST observation, as in an ODA procedure; it translates to surface wind and subsurface ocean memory, and thus provides fully-coupled, self-consistent initial conditions.

5. Toward Application to CGCM

So far we have only discussed application of CDA to relatively simple ICM and HCM systems, and have mostly used examples where nudging methods were the primary assimilation formalism. In a way, this is largely a reflection of what has been going on in this research area, but recently some preliminary work has been carried out on more advanced CDA methodology and toward its application to CGCMs. For instance, Cañizares *et al.* [24] developed a CDA procedure based on reduced-space Kalman Filters. The low-dimensional nature of this approach allowed them to determine via an inverse calculation (reduced space 4D-VAR for the Markov model approximating the full coupled model) a set of initial conditions which gives the best possible predictions. Such initial states, even when constrained to a 6-dimensional reduced space, showed forecast skill that outperforms the nudging-based initializations. To set the stage for CGCM use of CDA, a group of scientists at LDEO and IRI have also developed a set of bias correction schemes for a number of CGCMs.

Under the auspices of NOAA's Climate Test Bed program, a new research project was recently launched to study the predictability of ENSO and drought, with a focus on the generation and evaluation of long-term hindcasts using the NCEP Climate Forecast System (CFS), a state-of-the-art CGCM [25]. The main idea is to perform hindcast experiments with the CFS for the past one and a half centuries using reconstructed SST and sea level pressure (SLP) data, similar to what has been done with the LDEO model. However, like most of other CGCM forecast systems, the CFS is currently initialized in an uncoupled manner, which allows no feedbacks between the ocean and atmosphere during initialization. For both retrospective and operational forecasts, the initial conditions of the CFS are given by independently constructed oceanic (GODAS) and atmospheric (R2) reanalyses. This approach not only is prone to initialization shock, but also limits the length of hindcast experiments to the availability of the reanalysis datasets. Therefore, to achieve the objective of the project, application of CDA is a must. Admittedly, it is not an easy task to build a complete CDA system for CGCMs such as the CFS. For instance, one needs to construct a joint error covariance matrix for the ocean-atmosphere coupled system. A logical first step is to adopt the simple approach we have developed for the ICM. The main requirement here is that the CFS has realistic climatology and variability so that SST and SLP observations alone are enough to keep it on track of reality.

The CFS seems to meet the criterion nicely. Its ability in simulating the mean climate and ENSO was demonstrated in an analysis of 4 free runs of 32 years each [26]. The model climatology is quite realistic with relatively small systematic biases and climate drift, and the simulated tropical Pacific SST and wind anomalies are similar to those observed, as evident in Fig. 3. A set of retrospective forecasts with 15 ensemble members has also been produced with the CFS for the 24 year period of 1981–2004. It appears that the skill of the CFS in predicting ENSO is one of the best there is, though the skill assessment with this relatively short dataset is somewhat troublesome. The present study will overcome this shortcoming. In any case, since the CDA approach worked well in our ENSO ICM, there is no reason why it would not work for an advanced system like the CFS. With CDA implemented, the CFS is expected to do as well, if not better, in terms of ENSO simulation, and to go far beyond ENSO.

In order to develop a CDA procedure for the CFS, we first need to have a thorough understanding of the performance and biases of the system. This relies on analyses of the existing datasets of the CFS output, including free model runs and retrospective forecasts for the modern era. The essence of our procedure is a reduced space regression between observed SST/SLP fields and a selected set of model variables that represent the state of the coupled system, based on patterns identified by the Canonical Correlation Analysis (CCA) [27, 28]. CCA is a three-parametric prediction technique, which first uses principal component analysis to reduce dimensions of predictor and predictand spaces and then uses the singular value decomposition of the predictor-predictand covariance matrix of principal components to reduce the rank of predicting operator. In practice, assimilation procedure takes two steps: first, using the predetermined CCA prediction matrix, observed SST/SLP are converted to a "realistic" coupled model state; second, the predicted model state is assimilated into the CFS through a weighted nudging method. The latter can be considered the simplest form of Kalman filters, with weights equal to the effective ratios of observational and model error variances.

The CDA procedure has been tested with a set of short runs in which a large parameter range is explored. As an example, Fig. 9 compares CDA model runs using two different SST nudging parameters to observation and free model run. The equatorial upper ocean heat content anomaly in the CDA cases are similar to that observed and are much better than the case without data assimilation. Through the use of CDA, the information in the SST data has been translated to the dynamical ocean memory. The next step is to carry out a long initialization run with the CFS for the period from 1856 to present using CDA. This is like a free model run with the system being kept on track of reality by SST and SLP observations. This can be considered as an AMIP type experiment with an active ocean attached, or a CMIP type experiment with data assimilation. The resulting dataset will be used as initial condition as well as verification

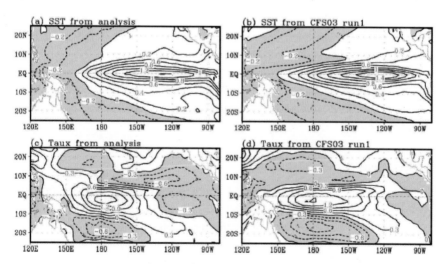

Fig. 8.   Regression against NINO3.4 SST. (a) SST from GODAS; (b) SST from CFS; (c) zonal momentum flux (Taux) from R2; (d) Taux from CFS. Units are dimensionless for (a) and (b), and 0.01 Pa/K for (c) and (d). GODAS and R2 datasets may be taken as observation. Adopted from [26].

1981-85   Euqatorial Anomalous Heat Content (0-300 m)

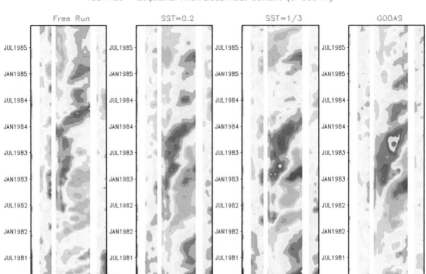

Fig. 9. Upper ocean (300 m) heat content anomalies along the equator as a function of time. The left panel is from a model run without data assimilation, the right panel is from observation, and the two middle panels are from model runs with CDA using different parameters for SST assimilation.

for long-term retrospective forecast experiment. This CDA dataset is basically a coupled reanalysis that will be very useful by its own right. For instance, we can use it to investigate the mechanisms, especially the oceanic processes, that are responsible for decadal and interdecadal climate variations, and to examine the interplay between these long-term variations and ENSO.

6. Summary and Conclusion

In this brief review, we first tried to demonstrate the importance and necessity of CDA for ENSO prediction by walking through the evolution history of LDEO model, and then discussed our current attempts toward application of CDA to CGCMs. In summary, we make the following concluding remarks:

- To prevent initialization shock and rapid climate drift, and thus to ensure a smooth and balanced forecast start, we need to develop and implement suitable CDA methodologies to initialize ocean-atmosphere coupled models for ENSO prediction.

- To study the predictability and the interdecadal modulation of ENSO, we need to create coupled reanalysis datasets and carry out retrospective forecast experiments over a long period of time with limited historical data, which is only feasible with CDA.

- So far there has been very limited work on CDA, which is out of proportion to the tremendous efforts and resources that have been devoted to ENSO research and prediction in the past few decades. More attention is needed in this research direction.

- Our current ability to predict ENSO is still far from realizing its potential. CDA may hold a key for further improvement of the predictive skill of present models. A prerequisite for successful CDA is an effective correction of systematic model biases.

Acknowledgments

This work is supported by research grants from National Basic Research Program (2007CB816005), National Science Foundation of China (40730843), National Oceanic and Atmospheric Administration, and National Aeronautics and Space Administration. The author would like to thank Dr. Wanqiu Wang for providing Figs. 8 and 9.

References

1. A. G. Barnston, A. Kumar, L. Goddard and M. P. Hoerling, *Bull. Am. Meteor. Soc.* **86** (2005) 59.
2. D. Chen, *Acta Oceanologica Sinica* **27** (2008) 1.
3. M. Latif *et al.*, *Climate Dyn.* **18** (2001) 255.
4. D. Chen and M. A. Cane, *J. Comput. Phys.* **227** (2008) 3625.
5. M. J. McPhaden *et al.*, *J. Geophys. Res.* **103** (1998) 14169.
6. D. Chen, *Int. J. Rem. Sens.* **22** (2001) 2621.
7. M. Latif *et al.*, *J. Geophys. Res.* **103** (1998) 14375.
8. D. Chen, *Sea Technol.* **38** (1997) 37.
9. D. Chen, S. E. Zebiak, A. J. Busalacchi and M. A. Cane, *Science* **269** (1995) 1699.
10. T. Lee, J.-P. Boulanger, A. Foo, L.-L. Fu and R. Giering, *J. Geophys. Res.* **105** (2000) 26063.
11. D. Chen, M. A. Cane, A. Kaplan, S. E. Zebiak and D. Huang, *Nature* **428** (2004) 733.

12. M. Rienecker, *Workshop Report: Coupled Data Assimilation Workshop*, *sponsored by NOAA/OGP* (2003), p. 23.
13. M. A. Cane, S. E. Zebiak and S. C. Dolan, *Nature* **321** (1986) 827.
14. S. E. Zebiak and M. A. Cane, *Mon. Weather Rev.* **115** (1987) 2262.
15. D. Chen, S. E. Zebiak, A. J. Busalacchi and M. A. Cane, *Mon. Weather Rev.* **125** (1997) 773.
16. D. Chen, M. A. Cane, S. E. Zebiak and A. Kaplan, *Geophys. Res. Lett.* **25** (1998) 2837.
17. D. Chen, M. A. Cane and S. E. Zebiak, *J. Geophys. Res.* **104** (1999) 11321.
18. D. Chen, M. A. Cane, S. E. Zebiak, R. Canizares and A. Kaplan, *Geophys. Res. Lett.* **27** (2000) 2585.
19. Y. Tang, Z. Deng, X. Zhou, Y. Cheng and D. Chen, *J. Climate* **21** (2008) 4811.
20. Y. Cheng, Y. Tang, X. Zhou, P. Jackson and D. Chen, *Climate Dyn.* (2009), doi:10.1007/s00382-009-0595-7.
21. A. Kaplan *et al.*, *J. Geophys. Res.* **103** (1998) 18567.
22. W. S. Kessler, *J. Climate* **14** (2001) 3055.
23. A. M. Moore and R. Kleeman, *J. Climate* **12** (1999) 1199.
24. R. Cañizares, A. Kaplan, M. A. Cane, D. Chen and S. E. Zebiak, *J. Geophys. Res.* **106** (2001) 30947.
25. S. Saha *et al.*, *J. Climate* **19** (2006) 3483.
26. W. Wang, S. Saha, H.-L. Pan, S. Nadiga and G. White, *Mon. Weather Rev.* **133** (2005) 1574.
27. T. P. Barnett and R. Preisendorfer, *Mon. Weather Rev.* **115** (1987) 1825.
28. A. G. Barnston and C. F. Ropelewski, *J. Climate* **5** (1992) 1316.

Advances in Geosciences
Vol. 18: Ocean Science (2008)
Eds. Jianping Gan et al.
© World Scientific Publishing Company

A NON-BOUSSINESQ TERRAIN-FOLLOWING
OGCM FOR OCEANOGRAPHIC
AND GEODETIC APPLICATIONS

Y. TONY SONG*, RICHARD GROSS, XIAOCHUN WANG,
and VICTOR ZLOTNICKI

*Jet Propulsion Laboratory, California Institute of Technology,
Pasadena, California 91106, USA*
* Tony.Song@jpl.nasa.gov

Altimetry sea-surface-height (SSH) represents ocean-volume changes, while gravimetry ocean-bottom-pressure (OBP) gives the vertically-integrated ocean-mass changes; both are fundamentally important information about ocean dynamics. Their proper representation in ocean general circulation models (OGCM) is essential for both studying ocean circulations and performing data assimilations by using the satellite data, as well as for geodetic applications. Here we show how these two oceanic parameters can be properly represented by a non-Boussinesq (mass-conserving) terrain-following OGCM. The innovative feature of the model is the stretched-pressure coordinate (sp-coordinate) system that generalizes the conventional sigma-coordinates and the s-coordinates without additional computational cost. In the new formulation, the OBP is the prognostic variable, while the SSH is a diagnostic variable, derived from the OBP with including the thermal expansion physics; therefore, they can be directly compared with the satellite data. This paper focuses on describing the model formulation, comparing with altimetry SSH and gravimetry OBP data, and demonstrating its capabilities for computing bottom pressure torques properly.

1. Introduction

The oceans cover 71% of the Earth's surface and their water masses exert a profound influence on our global climate. Satellite radar altimetry has been routinely observing ocean surface and its variability with a few centimeters accuracy and estimating global mean sea level to an accuracy of less than mm/year. The most important altimeters for sea level studies are those of TOPEX/Jason class that provide a near global map of sea surface height (SSH) from 1992 to present days in a time interval of few days and even daily by combining several altimeters (Fu and Cazenave, 2001). These altimetry data are the most important information of the ocean's volume changes because they contain changes in the temperature and salinity due to thermal expansion or contraction, as well as fresh water fluxes.

The main objectives of geodesy are the determination of the Earth's mean and time varying geometric shape, gravity informational field, rotation, and terrestrial reference frame. The variations in Earth rotation reflect both mass transport in the Earth system and the exchange of angular momentum among its components, which directly relate to changes in the ocean and its exchanges with other components (Gross et al., 2003). The torque of ocean bottom pressure (OBP) against topography has been theorized to play a key role in balancing the momentum imparted by the wind to the deep ocean circulation in the Antarctic Circumpolar Current (Munk and Palmen, 1951) and in the subtropical gyres (Hughes and de Cuevas, 2001). The oceanic mass changes are the immediate consequences of climate change, closely related to Earth's rotational variations and oceanic angular momentum budget on timescales from minutes to decades (Cox and Chao, 2002; Gross, 2007). Historically, the inability to represent bottom topography and mass-conservation in ocean general circulation models (OGCM) was one of the great obstacles to studying ocean-solid Earth interactions and assimilating OBP data (Condi and Wunsch, 2004). Model skill at estimating bottom pressure is of fundamental importance because the more accurate the model, the more effective will be data constraints on it (Fukumori et al., 1999).

To understand the Earth's mass changes, the U.S.-German Gravity Recovery and Climate Experiment (GRACE) has launched twin-satellites to measure the static and time varying components of Earth's gravity field (Tabley et al., 2004). The gravity field observed from space is largely a consequence of the mass distribution within the Earth system, which therefore can be used to derive the ocean bottom pressure change (Wahr et al., 1998; Hughes et al., 2002). The GRACE mission complements the existing altimeter satellites, such as the TOPEX/Poseidon (T/P) and Jason, that measure the sea surface height. In a homogeneous hydrostatic ocean, sea surface and bottom pressure variations are identical. In a stratified ocean, the two can be very different. The combination of GRACE data, representing oceanic mass changes, and altimeter data, representing oceanic volume changes, is very powerful (Jayne et al., 2003).

The key to successful application of GRACE and T/P-Jason data for both oceanography and solid-Earth studies is the model's capabilities in representing bottom topography and mass-conserving properties, which have not been addressed properly in currently used OGCMs (Condi and Wunsch, 2004).

The study of oceanic bottom pressure/mass redistribution clearly involves the issue of compressibility of seawater; therefore, the assumptions of conserving mass or volume in ocean models cannot be ignored (Huang and Jin, 2002). Our planet, as a whole, conserves its mass, but the distribution of mass within the Earth system, especially the hydrological components, is neither uniform nor constant in time (e.g. Chen et al., 2000). Sea level changes are not uniform globally and embody many aspects of oceanic density change, the global hydrological cycle, reflecting the heat content of oceans because the density of sea water depends on temperature (thermal expansion). The role played by the steric effect and oceanic mass redistribution in the observed global sea-level rise is still a major puzzle in current climate research (Douglas and Peltier, 2002). These fundamentally important issues have motivated us to construct a mass-conserving ocean model.

Most existing ocean models (Haidvogel and Beckmann, 1999; Griffies et al., 2000) are based on the incompressible approximations of seawater, or Boussinesq approximations (Boussinesq, 1903). These models calculate ocean flow while conserving water volume, instead of water mass, and sea-surface elevation is a prognostic variable, instead of ocean bottom pressure. These approximations to the basic equations of motion ignore the heat expansion/contraction physics that represent the real ocean (Huang and Jin, 2002) and are inconsistent with either T/P-Jason or GRACE data, unless ad-hoc (globally uniform) corrections are introduced (Greatbatch, 1994; Ponte, 1999). Even if a global uniform correction is applied, it is no guarantee against artificial changes in local or total mass of the oceans, unrelated to any real oceanographic effect (Condi and Wunsch, 2004). Recently, several publications have discussed removing the Boussinesq approximations in existing ocean models (e.g. DeSzoeke and Samelson, 2002; Song and Hou, 2004). Particularly, Losch et al. (2004) argues that the non-Boussinesq effects are most likely negligible with respect to other model uncertainties. However, their conclusions were based on a very coarse-resolution model (2-degree) without the terrain-following feature, which is the focus of this study. In addition, model uncertainties have been greatly reduced recently with increasing computational power that allows 1/10-degree or higher resolution models (Maltrud and McClean, 2005). Therefore, model's numerical errors are reducing; while the Boussinesq errors would not, and should be avoided if possible.

Another issue related to the study of ocean bottom pressure is topographic representation in ocean models. In practice, properly representing bottom topography in ocean models is a challenging task because it varies by the same order of the water depth. The topography in early ocean models had been approximated step-like by fixed vertical levels (z-coordinate system), like in the most widely used Bryan-Cox model. This vertical system has advantages in representing surface mixing and upper ocean dynamics because vertical levels can be easily adjusted. Its disadvantages have also been noticed, particularly in representing bottom kinematic conditions (Gerdes, 1993) and exchanging masses between shallow and deep waters (Roberts and Wood, 1997). Recently, progress has been made to overcome the topography problem by partial or shaved-cell technique (Adcroft et al., 1997).

However, we are more concerned with the representation of bottom pressure torques (Bell, 1999). Since ocean gyres are largely contained by bottom topography, the topographic issue in ocean models deserves a careful examination (Song and Haidvogel, 1994; Hughes and Cuevas, 2001; Song and Hou, 2006). For instance, in the z-coordinate models, only a number of ocean depths are permitted, the topography takes the form of a series of terraces, and there are many grid points at which the depth gradient is zero except the partial cell method is applied. In fact, many of the current studies did not apply the partial cell method (e.g. Maltrud and McClean, 2005) because it is not necessarily a simple option in many existing z-models and data-assimilation models. Since the bottom pressure torque is nonzero only where there is a slope, the terraces introduce artificially high viscous and nonlinear forces into the dynamics, as noticed by Hughes and de Cuevas (2001). Modeling study by Bryan (1997) has also showed that the smoothness of bottom topography had a large effect on the annual variability, while variations in isopycnal diffusivity and viscosity had much smaller effects on the motion of the axial oceanic angular momentum (OAM).

Previous studies have demonstrated that the hydrostatic, Boussinesq equation of motion in z-coordinates have the same form as the hydrostatic, non-Boussinesq equation in pressure coordinates (DeSzoeke and Samelson, 2002; Losch et al., 2004). However, such an issue has not been resolved in the generalized terrain-following coordinate models. Non-Boussinesq models in σ-coordinates have also been tested (Mellor and Ezer, 1995; Huang and Jin, 2002), but they are idealized cases and have never been compared with or evaluated by GRACE observations. Differing from these previous studies, we will focus on developing an innovative non-Boussinesq terrain-following

OGCM based on a stretched-pressure coordinate system. With the new system, these two problems — Boussinesq approximation and topography representation — can be resolved in a united form for both oceanographic and geodetic applications. We also show that OBP is the prognostic variable and therefore it can be directly compared with GRACE. The SSH can be diagnosed from the model OBP with including the thermal expansion and contraction physics, and it can be directly compared with altimeters.

The paper is arranged as follows: In Sec. 2, we describe our model formulation, particularly, issues on the *sp*-coordinate system and its representation of altimetry SSH and gravimetry OBP. In Sec. 3, we present the model results. Section 4 discusses the calculation of bottom pressure torque. Section 5 focuses on comparing the model with GRACE and altimetry data for the North Pacific. Conclusions are given in Sec. 6.

2. Model Formulation

2.1. *Basic equations*

We begin with the basic ocean equations in the Cartesian coordinate system with the z-axis pointing vertically upwards and the (x, y)-plane occupying the undisturbed water surface. The horizontal momentum equations are written in the form:

$$\frac{D\mathbf{v}}{Dt} + \mathbf{f} \times \mathbf{v} + \frac{1}{\rho}\nabla_z p = \frac{1}{\rho}\frac{\partial}{\partial z}\left(\rho K_M \frac{\partial \mathbf{v}}{\partial z}\right) + \mathbf{F}_M. \tag{1}$$

The tracer (temperature, salinity, biological, or chemical component) equations can likewise be written:

$$\frac{D\hat{T}}{Dt} = \frac{1}{\rho}\frac{\partial}{\partial z}\left(\rho K_H \frac{\partial \hat{T}}{\partial z}\right) + F_{\hat{T}}. \tag{2}$$

The hydrostatic balance equation is:

$$\frac{\partial p}{\partial z} + g\rho(T, S, p) = 0. \tag{3}$$

The mass continuity equation is:

$$\frac{1}{\rho}\frac{D\rho}{Dt} + \frac{\partial u}{\partial x} + \frac{\partial v}{\partial y} + \frac{\partial w}{\partial z} = 0, \tag{4}$$

where

$$\frac{D*}{Dt} = \frac{\partial *}{\partial t} + \mathbf{v} \cdot \nabla_z * + w\frac{\partial}{\partial z} *. \tag{5}$$

Notations used in these equations are as follows:

$$\mathbf{v} = (u, v) \quad \text{horizontal velocity}$$
$$w \quad \text{virtical velocity}$$
$$\hat{T}(x, y, z, t) \quad \text{tracers (temperature, salinity, etc.)}$$
$$\nabla_z = \left(\frac{\partial}{\partial x}, \frac{\partial}{\partial y}\right) \quad \text{horizontal gradient operator}$$
$$\rho = \rho_0 + \rho'(x, y, z, t) \quad \text{total density}$$
$$p \quad \text{pressure}$$
$$f = 2\Omega \sin\theta \quad \text{Coriolis due to Earth's rotation}$$
$$g \quad \text{acceleration due to gravity}$$
$$K_M(x, y, z, t) \quad \text{vertical eddy viscosity}$$
$$K_H(x, y, z, t) \quad \text{vertical eddy diffusivity}$$
$$\mathbf{F}_M, F_{\hat{T}} \quad \text{forcing, source, and viscous terms}$$

Notice that we have used $\hat{T}$ to represent a tracer component, including temperature T and salinity S. Discussions on the vertical eddy coefficients, horizontal viscous, and diffusive terms are referred to Song and Haidvogel (1994). In this paper the K-Profile Parametrization (KPP) scheme of Large et al. (1994) is used for vertical mixing and rotated tensors of Laplacian formulation are used for horizontal mixing.

It should be noted that the mass-conserving (non-Boussinesq) model can be easily simplified to the volume-conserving (Boussinesq) model (1) replacing the density ρ by the mean density ρ_0 in the momentum and tracers equations, (2) using geopotential depth to replace the pressure in the equation of state, i.e. $\rho(T, S, p) \approx \rho(T, S, p_0(z))$, and (3) using volume conservation in the mass continuity equation. However, the current non-Boussinesq formulation does not resolve the acoustic waves. This is because in the equation of state, $\rho = \rho(T, S, \bar{p})$, the density is still calculated from the salinity, potential temperature, and pressure, but the pressure is calculated from the hydrostatic relation. Therefore sound waves are filtered out in both the Boussinesq and the non-Boussinesq models (Mellor and Ezer, 1995).

2.2. The stretched-pressure coordinates

Before introducing our new coordinate system, we must define the ocean bottom pressure and its relation to the sea level. The pressure at the ocean

bottom is the sum of the atmospheric pressure and the weight per unit area
of the water column:

$$p_b = p_a + \int_{-h}^{\eta^*} g\rho dz, \tag{6}$$

where p_a is the atmospheric pressure, η^* is the sea level (the deviation
from the geoid height), and h is the bathymetry. Because an increase in p_a
of 1 mbar corresponds to a sea surface depression of approximately 1 cm,
the barometric correction to the sea level or an *inverted barometer* can be
written as

$$\eta^{ib} = -\frac{p_a}{g\rho_0}, \tag{7}$$

where ρ_0 is the mean density. With this definition, the ocean bottom
pressure can be further written as

$$p_b \approx \int_{-h}^{\eta^* - \eta^{ib}} g\rho dz = \int_{-h}^{\eta} g\rho dz. \tag{8}$$

Here we have defined the *barometrically corrected sea level* as

$$\eta = \eta^* - \eta^{ib}. \tag{9}$$

These conventions are consistent with both altimetry SSH and gravimetry
OBP data. Most gridded altimetry SSH data have the IB removed
(the along track data maybe not). The GRACE OBP is the sum of
the atmospheric pressure and the vertical integral of the water mass,
which includes IB (Chambers, 2008). The conventions also simplify model
numerics by only applying the atmospheric pressure to pressure gradient
terms; therefore, their dynamics effect to the ocean has also been included
(Ponte, 1999).

To solve the non-Boussinesq ocean equations without losing the
topography-following feature, we introduce a new non-linear pressure-
coordinate system, called stretched pressure coordinates (or *sp-*
coordinates for short), as

$$p = -(p_b' + p_c)s - (p_b^0 - p_c)C(s), \tag{10}$$

where $-1 \leq s \leq 0$ with $s = 0$ on the ocean surface and $s = -1$ at the
ocean bottom, $p_b^0(x, y) = p_b(x, y, 0)$ is the initial or mean bottom pressure,
$p_b' = p_b(x, y, t) - p_b^0(x, y)$ is the bottom pressure anomaly, p_c is a constant
pressure near the thermocline or the shallowest depth, and $C(s)$ is the

stretching function of Song and Haidvogel (1994), which has a property of $C(0) = 0$ and $C(-1) = -1$, as detailed in the appendix. It should be noted that the current system is nonlinear, which has advantages over the traditional pressure or σ-coordinates used in atmospheric or oceanic models (Huang and Jin, 2002; Mellor and Ezer, 1995), allows enhanced resolutions near the surface and bottom layers of the ocean, and recovers the traditional σ-coordinates without additional cost.

Using the hydrostatic equation, we obtain

$$g\rho\frac{\partial z}{\partial s} = (p'_b + p_c) + (p^0_b - p_c)C'(s). \tag{11}$$

For convenience, we introduce two physical parameters: **the metric parameter** $H_z = \frac{\partial z}{\partial s}$ and **the Boussinesq parameter** $B_z = \frac{\rho}{\rho_0}$, and define their product as **parametric function**

$$\phi(x, y, s, t) = H_z(x, y, s, t)B_z(x, y, s, t). \tag{12}$$

These equations are a particular case of the parametric vertical coordinate formulation of Song and Hou (2006).

Using the relationships (10)–(12) and following Song (1998) and Song and Hou (2006), the generalized coordinate Eqs. (1)–(5) can be written

$$\frac{\partial(\phi u)}{\partial t} + \frac{\partial(\phi uu)}{\partial x} + \frac{\partial(\phi vu)}{\partial y} + \frac{\partial(\phi \Omega u)}{\partial s} - \phi fv$$

$$= -\frac{\phi}{B_z}\left\{\frac{1}{\rho_0}\frac{\partial p_a}{\partial x} + g\frac{\partial \eta}{\partial x} - \int_s^0 \left(\frac{\partial z}{\partial s'}\frac{\partial b'}{\partial x} - \frac{\partial z}{\partial x}\frac{\partial b'}{\partial s'}\right)ds'\right\}$$

$$+ \frac{\partial}{\partial s}\left(\frac{B_z^2 K_M}{\phi}\frac{\partial u}{\partial s}\right) + \phi \mathcal{F}_u \tag{13}$$

$$\frac{\partial(\phi v)}{\partial t} + \frac{\partial(\phi uv)}{\partial x} + \frac{\partial(\phi vv)}{\partial y} + \frac{\partial(\phi \Omega v)}{\partial s} + \phi fu$$

$$= -\frac{\phi}{B_z}\left\{\frac{1}{\rho_0}\frac{\partial p_a}{\partial y} + g\frac{\partial \eta}{\partial y} - \int_s^0 \left(\frac{\partial z}{\partial s'}\frac{\partial b'}{\partial y} - \frac{\partial z}{\partial y}\frac{\partial b'}{\partial s'}\right)ds'\right\}$$

$$+ \frac{\partial}{\partial s}\left(\frac{B_z^2 K_M}{\phi}\frac{\partial v}{\partial s}\right) + \phi \mathcal{F}_v \tag{14}$$

$$\frac{\partial(\phi \hat{T})}{\partial t} + \frac{\partial(\phi u \hat{T})}{\partial x} + \frac{\partial(\phi v \hat{T})}{\partial y} + \frac{\partial(\phi \Omega \hat{T})}{\partial s}$$

$$= \frac{\partial}{\partial s}\left(\frac{B_z^2 K_H}{\phi}\frac{\partial \hat{T}}{\partial s}\right) + \phi \mathcal{F}_{\hat{T}} \tag{15}$$

and

$$\frac{\partial(\phi)}{\partial t} + \frac{\partial(\phi u)}{\partial x} + \frac{\partial(\phi v)}{\partial y} + \frac{\partial(\phi \Omega)}{\partial s} = 0 \tag{16}$$

where Ω is the vertical velocity in the new coordinate system, defined by

$$\Omega = \frac{\partial s}{\partial z}\left\{ w - \left(\frac{\partial z}{\partial t}\right)_s - u\frac{\partial z}{\partial x} - v\frac{\partial z}{\partial y} \right\}, \tag{17}$$

$b' = -g\rho'/\rho_0$ is the buoyancy, the vertical high can be obtained by

$$z = \int_{-1}^{s} \frac{\phi}{B_z} ds' - h. \tag{18}$$

It is important to note that assuming $B_z = 1$ in Eqs. (13)–(17) recovers the conventional Boussinesq primitive equations. In addition, these equations are subject to boundary conditions on sea surface $s = 0$ and at ocean bottom $s = -1$.

The surface conditions, evaluated at $z = \eta$ (or $s = 0$), are

$$\frac{B_z^2 K_M}{\phi}\frac{\partial u}{\partial s} = \tau_s^x(x, y, t), \tag{19}$$

$$\frac{B_z^2 K_M}{\phi}\frac{\partial v}{\partial s} = \tau_s^y(x, y, t), \tag{20}$$

$$\frac{B_z^2 K_H}{\phi}\frac{\partial T}{\partial s} = \frac{Q_T}{\rho_0 C_p} + \frac{1}{\rho_0 C_p}\frac{dQ_T}{dT}(T - T_{\text{ref}}), \tag{21}$$

$$\frac{B_z^2 K_H}{\phi}\frac{\partial S}{\partial s} = \frac{\rho_f}{\rho_0}q_w S, \tag{22}$$

$$\phi\Omega = -\frac{\rho_f}{\rho_0}q_w, \tag{23}$$

where τ_s^x and τ_s^y are the components of wind stress acting on the free surface in the x and y directions, respectively, Q_T is the heat fluxes, C_p is the heat capacity of sea water, ρ_f is the density of fresh water, and q_w is the net fresh water flux into the ocean (precipitation plus river runoff minus evaporation). Here T and S are the temperature and salinity, respectively. Boundary conditions for other tracers have to be applied accordingly.

Noted that the surface conditions of the salinity and the vertical velocity, Ω, in the sp-coordinates differ from the conventional Boussinesq models. In this non-Boussinesq formulation, the natural boundary condition for salinity balance is used, i.e. the advective flux of salinity cancels the turbulent flux term and there is no salt flux through the air-sea interface.

The fresh water flux is directly applied to the vertical velocity, which is converted into ocean bottom pressure in the barotropic equations:

$$\frac{\partial p_b}{\partial t} + \nabla_s \left\{ \int_{-1}^{0} (\phi u, \phi v) ds \right\} = \frac{\rho_f}{\rho_0} q_w. \tag{24}$$

Correspondingly, at the sea bed, $z = -h$ (or $s = -1$), the boundary conditions are

$$\frac{B_z^2 K_M}{\phi} \frac{\partial u}{\partial s} = \tau_b^x(x, y, t), \tag{25}$$

$$\frac{B_z^2 K_M}{\phi} \frac{\partial v}{\partial s} = \tau_b^y(x, y, t), \tag{26}$$

$$\frac{B_z^2 K_H}{\phi} \frac{\partial T}{\partial s} = 0, \tag{27}$$

$$\frac{B_z^2 K_H}{\phi} \frac{\partial S}{\partial s} = 0, \tag{28}$$

$$\phi \, \Omega = 0, \tag{29}$$

where $\tau_b^x = (\gamma_1 + \gamma_2\sqrt{u^2 + v^2})u$ and $\tau_b^y = (\gamma_1 + \gamma_2\sqrt{u^2 + v^2})v$, and γ_1 and γ_2 are coefficients of linear and quadratic bottom friction, respectively. It should be noted that this combined linear and quadratic formulation of the bottom drag has the flexibility for both shallow or deep ocean applications. For deep ocean, if only the linear bottom drag is needed, γ_2 can be set to zero. Similarly, γ_1 can be set to zero for shallow seas.

In the non-Boussinesq formulation, the OBP is a prognostic variable, i.e., the bottom level of the pressure, which can be derived from (10) by setting $s = -1$:

$$p_b(x, y, t) = p_b^0 + p_b'(x, y, t). \tag{30}$$

The free surface becomes a diagnostic variable, derived from (18) by setting $s = 0$:

$$\eta = \int_{-1}^{0} \frac{\phi}{B_z} ds' - h. \tag{31}$$

With this new formulation, the thermal expansion and contraction are included in deriving the sea surface elevation because of the Boussinesq parameter B_z in the equation. Setting $B_z = 1$ recovers the conventional Boussinesq model. It should be noted that the sp-coordinate system is a particular case of the parametric coordinates of Song and Hou (2006), therefore the numerical methods for solving the equations are the same and will not be repeated here.

3. The Non-Boussinesq Global Ocean

The purpose of this section is to demonstrate the capability of the new model in handling global-scale applications for long-term simulations with both non-Boussinesq and terrain-following features. The global model covers the world ocean from 75°S to 75°N with realistic topography derived from the ETOPO5 database. The model grid resolution is 0.5 × 0.5 degree, with enhanced resolution to 1/3° in the range of 30°S to 30°N. Initial T and S are obtained from the Levitus et $al.$ (1994) climatology. Surface boundary conditions include heat and fresh water fluxes and wind stress obtained from NCEP/NCAR reanalysis data. There are 20 vertical levels with 10 levels in the upper 500-meter depth and the other 10 levels in the lower layer, using the sp-coordinate system of formulation (10).

The model is first spun-up for 50 years with annual-mean NCEP/NCAR forcing. Time series of global averaged kinetic energy and potential energy are shown in Fig. 1 to demonstrate the extent to which a quasi-steady state has been reached at the end of the 50-year spun-up run. The early 10 years of the run are dominated by a kind of geostrophic adjustment process in which the potential energy convert rapidly to the kinetic energy, which is then subject to dissipation. The physical reason for the energy conversion has been discussed by Song and Wright (1998) as the advective elimination of the inconsistency between the finite-difference nature of the model and the initial state. Such an adjustment process is required for energetic consistency and the success of long-term integrations. However, the dissipation process is extremely slow, as it can be seen that the potential energy still increases slightly after 30 years. This is related to the values of horizontal viscosity and diffusivity used in the model, which are 400 and $100\,\mathrm{m}^2\,\mathrm{s}^{-1}$, respectively.

After the 50-year spun-up run, the model is then integrated from year 1950 to year 2006 with the monthly-mean NCEP/NCAR forcing. Figures 2 and 3 compare the GRACE-inferred OBP with the model OBP in annual amplitude and phase, respectively. Their amplitudes of the variability are about 4 mbar. Clearly, there are big differences in both the annual amplitude and phase changes between GRACE and the model, particularly along the coastline. This land signal contamination is actually known to the GRACE community, which is due the smoothing filters used in the GRACE data

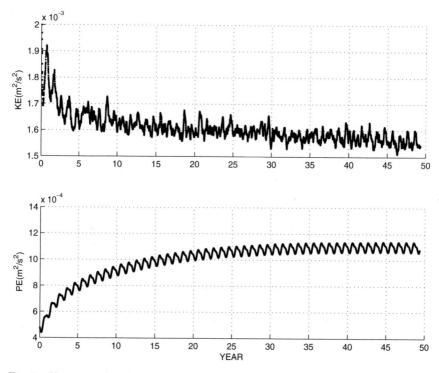

Fig. 1.   Upper panel is the globally averaged kinetic energy and the lower panel is the potential energy during a 50-year spun-up run.

processing (Wahr *et al.*, 2004; Chambers, 2008). Because the signal of mass change at the land side is larger than that at the ocean side, the smoothing would attenuate the amplitude at the land side. This attenuation represents a loss of a part of the signal at the land side that is added to the smoothed value at the ocean side, or leaked into the ocean side. In addition, earthquakes also cause the ocean floor and gravity field changes. The strong signal in the northeast corner of the Indian Ocean is mainly due to the 2004 Sumatra-Andaman earthquakes, which generated the deadly tsunami (Song *et al.*, 2008). Nevertheless, the model gives the first look of the ocean bottom pressure, which is reasonably similar to what the GRACE have been observing.

Another way to evaluate the model OBP is to compare the model SSH with altimeters. This is because the model SSH is diagnosed from the OBP and they have to be physically consistent. Similarly, Figs. 4 and 5 compare the altimetry SSH with the model in annual amplitude and phase,

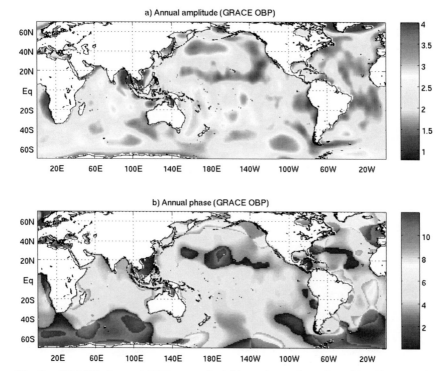

Fig. 2.   GRACE-observed OBP anomaly $(p'_b(x, y, t) = p_b(x, y, t) - p^0_b(x, y))$ from 2002 to 2006: upper panel is the annual amplitude in mbar and lower panel is the phase in month.

respectively. Their amplitudes are about 8 cm and their phases agree very well. The location of the strong variability in the Kuroshio and Gulf Stream suggests that the separation is occurring at about the correct latitude. The separation point has been a long problem in many OGCMs except in those very high-resolution models as reported in Maltrude and McClean (2005). The most striking signatures are the Antarctic Circumpolar Current, particularly in the OBP variability field. Our model results are reasonably well in comparison with known observations.

Perhaps this is the first time such a non-Boussinesq global model with terrain-following capability has been run for a 50-year simulation. This experiment illustrates that the model with the new scheme is stable for long-term integrations and capable of resolving global-scale problems with realistic topography. The SSH gives water-volume changes; while OBP represents ocean-mass changes. In a homogeneous hydrostatic ocean,

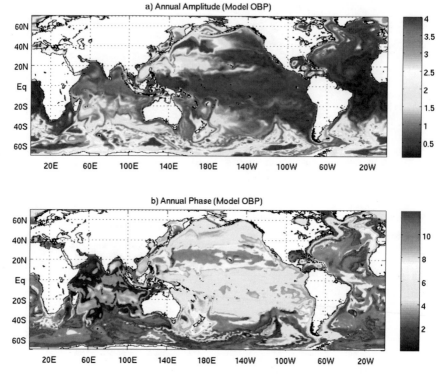

Fig. 3.   Same as Fig. 2, but from the model.

sea surface and bottom pressure variations are identical. In a stratified
ocean, the two can be very different. Here we show that these two oceanic
parameters can be properly represented by a non-Boussinesq (conserving
mass) terrain-following global ocean model, therefor they can be directly
compared with GRACE and altimeters for geodetic applications.

4. Ocean Bottom Pressure Torque

As mentioned in the introduction, bottom pressure torque is an quantity
measuring the interactions between ocean and the solid Earth and useful
for both oceanographic and geodetic applications. For example, Holland
(1973) noted that the action and reaction of pressure forces on topographic
slopes could lead to large bottom pressure torque, which reinforces the
effect of the wind-stress curl, resulting in an increased northward flow in the
Gulf Stream of the model. Wunsch and Roemmich (1985) also suggested

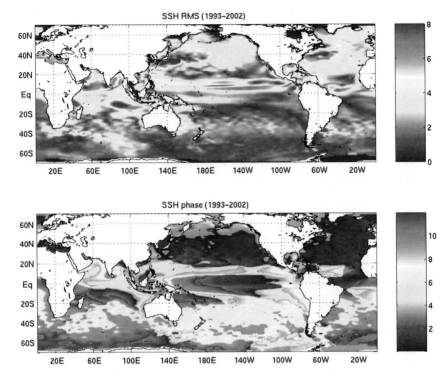

Fig. 4. Altimetry SSH from 1993 to 2006: upper panel is the annual amplitude in cm and lower panel is the phase in month.

that topographic interactions are necessary in explaining observations in the North Atlantic. In geodesy, oceanic mass redistribution and current velocity change are the major force in oceanic angular momentum and the Earth's rotational changes (Cox and Chao, 2002; Gross, 2007). These can be explained by the vorticity balance equation (with the rigid lid assumption at the surface for simplicity):

$$\beta \frac{\partial \Psi}{\partial x} = J(p_b, h) + curl(\tau - \tau_b), \qquad (32)$$

where Ψ is the streamfunction of the depth-integrated mass transport, τ is the wind stress, τ_b is the frictional stress of the flow on the bottom, $\beta = df/dy$, and the Jacobian, $J(a, b) = a_x b_y - a_y b_x$, is the bottom pressure torque. The bottom friction term is generally small, but the bottom pressure torque can be locally very large (Olbers et al., 2004).

78   Y. T. Song et al.

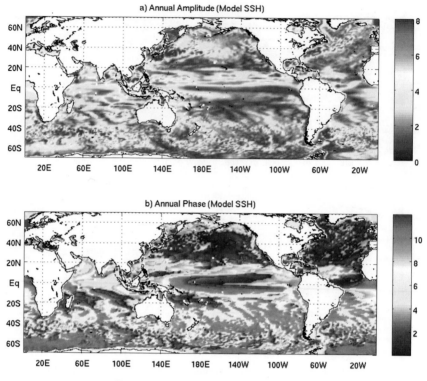

Fig. 5. Same as Fig. 4, but from model.

To better understand the vorticity balance of the ocean, the proper calculation of bottom pressure torque is essential. The torque of ocean bottom pressure against topography is believed to play a key role in balancing the momentum imparted by the wind to the deep ocean circulation in the Antarctic circumpolar current (Munk and Palmen, 1951) and in the subtropical region (Hughes and de Cuevas, 2001). For example, an outstanding feature of the Southern Ocean dynamics is the failure of one of the cornerstones of theoretical oceanography — the Sverdrup balance $\beta\frac{\partial\Psi}{\partial x} = curl(\tau)$. Clearly, in the range of latitudes of Drake Passage, the Sverdrup balance must fail because the circumpolar integral of the meridional mass transport $\frac{\partial\Psi}{\partial x}$ is zero to ensure mass conservation; while the wind stress curl will not integrate to zero in general. The reason for the failure of the Sverdrup theory is because it neglects the interaction of the circulation with topography, represented by the bottom pressure torque. Apparently, the same is true for most of the world ocean (Hughes and de Cuevas, 2001).

In designing ocean models, Song and Wright (1998) emphasized the numerical consistency of bottom pressure torque as important as the consistency of energy and momentum of the ocean. The combination of the terrain-following coordinates and the non-Boussinesq bottom pressure provides a framework to maintain the consistency because the bottom pressure is exactly computed at the topography location. In Fig. 6, we have

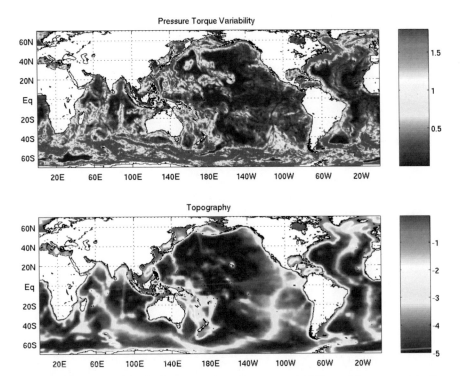

Fig. 6. Upper panel is the bottom pressure torque variability and lower panel is the topography. Unites are $10^{-6}\,\mathrm{Nm}^{-3}$ and kilometers, respectively. The torque of ocean bottom pressure against topography is believed to play the key role in balancing the momentum imparted by the wind to the deep ocean circulation in the Antarctic circumpolar current (Munk and Palmen, 1951) and in the subtropical region (Hughes and de Cuevas, 2001). The terrain-following feature of the model simplifies the calculation of the bottom pressure torque. It can be seen that its maximum variability is the results of ocean-solid Earth interactions, especially over the Antarctic circumpolar current and western boundary regions.

calculated the bottom pressure torque and the topography of the global ocean. The terrain-following feature of the model allows us to calculate the bottom pressure torque easily. It can be seen that its maximum variability is the result of the Antarctic circumpolar current and western boundary current imparted on the topography.

5. Focusing on the North Pacific

Because it is still difficult to evaluate the model globally by GRACE due to the leakage of the land hydrological signals to the ocean in the gravity coefficients, here we focus on the North Pacific signal that is away from the boundaries. Figure 7 validates the model by comparing the area-averaged OBP with GRACE and SSH with altimeters. The time-series are the averaged values in the subpolar area minus that in

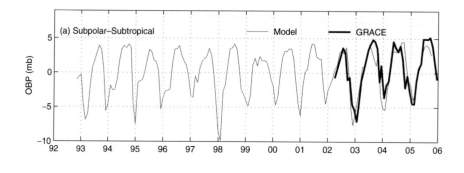

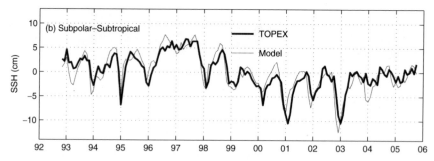

Fig. 7. Comparison of model with (a) GRACE-observed OBP and (b) altimetry SSH for the region of North Pacific. Time-series are the averaged values of the subpolar area minus the subtropical area. Year numbers indicate the beginning of the year.

the subtropical area. First, we can see that the model simulates the GRACE OBP faithfully for the observed period. Second, the model results agree well with altimetry SSH, indicating that the diagnosed sea surface is physically consistent with OBP. Since OBP and SSH have to be consistent in representing oceanic mass changes and volume changes, respectively, our non-Boussinesq model seems to represent the physics of these two variables well. In Figs. 8 and 9, we have compared the spatial patterns of the composite monthly mean OBP for the winter and fall, respectively. Again, the model does a good job in simulating the oscillation features in the North Pacific, as studied by Song and Zlotnicki (2008).

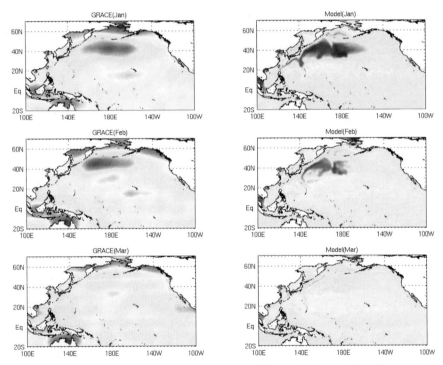

Fig. 8.   Monthly mean OBP for the winter season (January, February, March) composed from 2002 to 2006. Left panels from GRACE and right panels from the model. Red and orange colors indicate positive anomaly while blue and green colors indicate negative anomaly, range from −4 mb to +4 mb.

82   *Y. T. Song et al.*

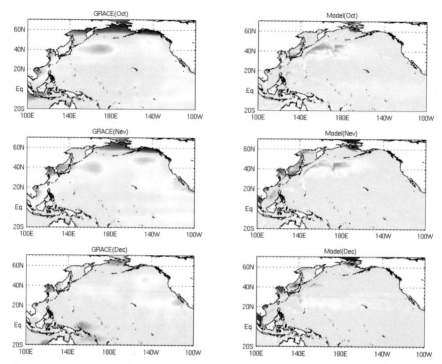

Fig. 9. Monthly mean OBP for the fall season (October, November, December) composed from 2002 to 2006. Left panels from GRACE and right panels from the model. Red and orange colors indicate positive anomaly while blue and green colors indicate negative anomaly, ranging from −4 mb to +4 mb.

6. Conclusions

The main purpose of this paper is to implement the *sp*-coordinate system into an ocean circulation model. Some of the science applications are shown in Song and Zlotnicki (2004, 2008) and Zlotnicki *et al.* (2007). More work needs to be done, particularly, in comparing this model with the conventional Boussinesq models. Nevertheless, we have successfully demonstrated that ocean bottom pressure and sea surface height can be properly represented in a non-Boussinesq terrain-following global ocean model. In such a model, the ocean mass is conserved and the heat expansion physics can be included in calculating these two variables, and therefore can be directly compared with and constrained by GRACE and T/P-Jason observations. The terrain-following capability allows easy calculation of ocean bottom pressure torques, helping applications of GRACE data for both oceanographic and geodetic studies.

Acknowledgments

The research described here was conducted at the Jet Propulsion Laboratory, California Institute of Technology, under contract with the National Aeronautics and Space Administration.

Appendix 1. The S-Coordinate System

The s-coordinate system of Song and Haidvogel (1994) is:

$$z = \zeta(1 + s) + h_c s + (h - h_c)C(s),$$

where $-1 \leq s \leq 0$ and $C(s)$ is a set of s-curves, defined by

$$C(s) = (1 - b)\frac{\sinh(\theta s)}{\sinh \theta} + b\frac{\tanh[\theta(s + \frac{1}{2})] - \tanh(\frac{1}{2}\theta)}{2\tanh(\frac{1}{2}\theta)}$$

where θ and b are the surface and bottom control parameters. Their ranges are $0 \leq \theta \leq 20$ and $0 \leq b \leq 1$, respectively. h_c is a constant chosen to be the minimum depth of the bathymetry or a width of surface or bottom boundary layer in which a higher resolution is required.

References

1. A. Adcroft, C. Hill and J. Marshall, Representation of topography by shaved cells in a height coordinate ocean model, *Mon. Weather Rev.* **125** (1997) 2293–2315.
2. J. M. Bell, Vortex stretching and bottom torques in the Bryan–Cox ocean circulation model, *J. Geophys. Res.* **104** (1999) 23545–23563.
3. J. Boussinesq, *Theorie Analyque de la Chaleur*, vol. 2 (Paris: Gathier-Villars, 1903).
4. F. O. Bryan, The axial angular momentum balance of a global ocean general circulation model, *Dyn. Atmos. Oceans* **25** (1997) 121–216.
5. D. P. Chambers, Converting release-04 gravity coefficients into maps of equivalent water thickness, Report, CSR, Univ. Texas, Austin (2008), p. 9.
6. J. L. Chen, C. K. Shum, C. R. Wilson, D. P. Chambers and B. D. Tapley, Seasonal sea level change from TOPEX/Poseidon observation and thermal contribution, *J. Geodesy.* **73** (2000) 638–647.
7. F. Condi and C. Wunsch, Measuring gravity field variability, the geoid, ocean bottom pressure fluctuations, and their dynamical implications, *J. Geophys. Res.*, vol. 109 (2004), doi:10.1029/2002JC001727.

8. C. M. Cox, and B. F. Chao, Detection of a large-scale mass redistribution in the terrestrial system since 1998, *Science* **297** (2002) 831–833.

9. B. C. Douglas and W. R. Peltier, The puzzle of global sea-level rise, *Physics Today* (2002) 35–40.

10. R. A. De Szoeke and R. M. Samelson, The duality between the Boussinesq and non-Boussinesq hydrostatic equations of motion, *J. Phys. Oceanogr.* **23** (2002) 2194–2203.

11. L.-L. Fu and A. Cazenave (eds.), *Satellite Altimetry and Earth Sciences* (Academic Press, 2001), p. 463.

12. I. Fukumori, R. Raghunath, L.-L. Fu and Y. Chao, Assimilation of TOPEX/Poseidon altimeter data into a global ocean circulation model: How good are the results? *J. Geophys. Res.* **104** (1999) 25647–25665.

13. R. Gerdes, A primitive equation ocean circulation model using a general vertical coordinate transformation, I. Description and testing of the model, *J. Geophys. Res.* **98** (1993) 14683–14701.

14. R. J. Greatbatch, A note on the representation of steric sea level in models that conserve volume rather than mass, *J. Geophys. Res.* **99** (1994) 12767–12772.

15. S. M. Griffies, C, Boning, F. O. Bryan, E. P. Chassignet, R. Gerdes, H. Hasumi, A. Hirst and A. M. Treguier, Developments in ocean climate modeling, *Ocean Model.* **2** (2000) 123–192.

16. R. S. Gross, Earth rotation variations long period, in *Physical Geodesy*, ed. T. A. Herring (Elsevier, Oxford, 2007), pp. 239–294.

17. R. S. Gross, I. Fukumori and D. Menemenlis, Atmospheric and oceanic excitation of the Earth's wobbles during 1980–2000, *J. Geophys. Res.*, vol. 108 (2003), Article No. 2370.

18. W. R. Holland, Baroclinic and topographic influences on the transport in western boundary currents, *Geophys. Fluid Dyn.* **4** (1973) 187–210.

19. D. B. Haidvogel and A. Beckmann, *Numerical Ocean Circulation Modeling* (Imperial College Press, 1999), p. 18.

20. R. X. Huang and X. Jin, Sea surface elevation and bottom pressure anomalies due to thermohaline forcing, Part I: Isolated perturbations, *J. Phys. Oceanogr.* **32** (2002) 2131–2150.

21. C. W. Hughes and B. A. de Cuevas, Why western boundary currents in realistic oceans are inviscid: A link between form stress and bottom pressure torques, *J. Phys. Oceanogr.* **31** (2001) 2871–2885.

22. C. W. Hughes, C. Wunsch and V. Zlotnicki, Satellite peers through the oceans from space. EOS, *Trans. Am. Geophys. Union* **81** (2002) 68.

23. S. R. Jayne, J. M. Wahr and F. Bryan, Observing ocean heat content using satellite gravity and altimetry, *J. Geophys. Res.* **108** (2003) 1029–1042.

24. W. G. Large, J. C. McWilliams and S. C. Doney, Oceanic vertical mixing: A review and a model with a nonlocal boundary layer parametrization, *Rev. Geophys.* **32** (1994) 363–403.

25. S. Levitus, R. Burget and T. Boyer, World ocean altas: Salinity and temperature 3–4, NOAA Altas NESDID, U.S. Dept. Commerce (1994).

26. M. Losch, A. Adcroft and J. Campin, How sensitive are coarse general circulation models to fundamental approximations in the equations of motion? *J. Phys. Oceanogr.* **34** (2004) 306–319.

27. M. E. Maltrud and J. L. McClean, An eddy resolving global 1/10 degree ocean simulation, *Ocean Model.* **8** (2005) 31–54.

28. G. L. Mellor and T. Ezer, Sea level variations induced by heating and cooling: An evaluation of the Boussinesq approximation in ocean models, *J. Geophys. Res.* **100** (1995) 20565–20577.

29. W. H. Munk and E. Palmen, Note on the dynamics of the Antarctic Circumpolar currents, *Tellus* **3** (1951) 53–55.

30. D. Olbers, D. Borowski, C. Volker and J. Wolff, The dynamical balance, transport and circulation of the Antarctic Circumpolar Current, *Antarctic Science* **16** (2004) 439–470.

31. R. Ponte, A preliminary model study of the large-scale seasonal cycle in bottom pressure over the global ocean, *J. Geophys. Res.* **104** (1999) 1289–1300.

32. M. J. Roberts and R. A. Wood, Topographic sensitivity studies with a Bryan–Cox–Type ocean model, *J. Phys. Oceanogr.* **27** (1997) 823–836.

33. Y. T. Song, A general pressure gradient formulation for ocean models. Part I: Scheme design and diagnostic analysis, *Mon. Weather Rev.* **126** (1998) 3213–3230.

34. Y. T. Song, L.-L. Fu, V. Zlotnicki, C. Ji, V. Hjorleifsdottir, C. K. Shum and Y. Yi, The role of horizontal impulses of the faulting continental slope in generating the 26 December 2004 Tsunami, *Ocean Model.* (2008), doi:10.1016/j.ocemod.2007.10.007.

35. Y. T. Song and D. B. Haidvogel, A semi-implicit ocean circulation model using a generalized topography-following coordinate, *J. Comput. Phys.* **115** (1994) 228–244.

36. Y. T. Song and T. Y. Hou, Parametric vertical coordinate formulation for multiscale, Boussinesq, and non-Boussinesq ocean modeling, *Ocean Model.* (2006), doi:10.1016/j.ocemod.2005.01.001.

37. Y. T. Song and D. Wright, A general pressure gradient formulation for ocean models, Part II: Energy, momentum, and bottom torque consistency, *Mon. Weather Rev.* **126** (1998) 3231–3247.

38. Y. T. Song and V. Zlotnicki, Ocean bottom pressure waves predicted in the tropical Pacific, *Geophys. Res. Lett.* **31** (2004) L05306, doi: 10.1029/2003GL018980.

39. Y. T. Song and V. Zlotnicki, Subpolar ocean-bottom-pressure oscillation and its links to the tropical ENSO, *Int. J. Rem. Sens.* **29** (2008) 6091–6107.

40. B. D. Tapley, S. Bettadpur, M. Watkins and C. Reigber, The gravity recovery and climate experiment; mission overview and early results, *Geophys. Res. Lett.* **31** (2004) L09607, doi: 10.1029/2004GL019920.

41. J. Wahr, M. Molenaar and F. Bryan, Time variability of the Earths gravity field: Hydrological and oceanic effects and their possible detection using GRACE, *J. Geophys. Res.* **103** (1998) 30205–30229.

42. J. Wahr, S. Swenson, V. Zlotnicki and I. Velicogna, Time-variable gravity from GRACE: First results, *Geophys. Res. Lett.* **31** (2004) L11501, doi:10.1029/2004GL019779.
43. C. Wunsch and D. Roemmich, Is the North Atlantic is Sverdrup balance? *J. Phys. Oceanogr.* **15** (1985) 1856–1880.
44. V. Zlotnicki, J. Wahr, I. Fukumori and Y. T. Song, The antarctic circumpolar current: Transport variability during 2002–2005, *J. Phys. Oceanogr.* **37** (2007) 231–244.

Advances in Geosciences
Vol. 18: Ocean Science (2008)
Eds. Jianping Gan et al.
© World Scientific Publishing Company

TROPICAL PACIFIC UPPER OCEAN HEAT CONTENT VARIATIONS AND ENSO PREDICTABILITY DURING THE PERIOD FROM 1881–2000

YOUMIN TANG* and ZIWANG DENG

Environmental Science and Engineering
University of Northern British Columbia,
Prince George, BC, Canada
*ytang@unbc.ca

In this study, a long-term analysis of the tropical Pacific upper ocean heat content (HC) was obtained for the period from 1881–2000, by assimilating historic sea surface temperature dataset into an oceanic general circulation model (OGCM) with Ensemble Kalman filter. The validation against the NCEP (National Center of Environmental Prediction) HC and the observed HC indicates that the analyzed HC captures very well the large-scale observed features of HC. There exists a striking interannual variability in the tropical Pacific upper ocean HC anomalies (HCA). Like ENSO (El Niño and the Southern Oscillation), the HCA interannual variability also has a significant interdecadal variation. The interdecadal variation in the HCA causes the interdecadal variation in the lagged correlation between the HCA of the equatorial western Pacific ocean and the SSTA (sea surface temperature anomalies) of the equatorial eastern Pacific, which in turn affects ENSO prediction skill (Niño3.4 SSTA). The long-term retrospective ENSO prediction from 1881–2000 by the model supported the above conclusion.

1. Introduction

The tropical Pacific upper hear content (HC) is an important component of the coupled ocean-atmosphere system of the tropical Pacific ocean on the interannual timescale, and a major source of memory for the system. It plays an essential role in the oscillation of the ENSO cycle by controlling the temperature of the waters upwelled in the eastern equatorial Pacific. The link of ENSO variability to the heat content build-up and discharge in the tropical Pacific has been evidenced in theory and observation (e.g. Wyrtki, 1985; Suarez and Schopf, 1988; Battisti, 1988; Jin, 1997). It has been found that the HC redistribution in the western tropical Pacific can lead to the evolution of SST anomalies in the eastern Pacific and has been known to be an important factor in the evolution of many ENSO episodes. In particular, the HC anomalies over the equatorial band 5°N to 5°S can be a very good precursor for the SST anomalies in the Niño3 region (5°N–5°S, 150°W–90°W) (e.g. Zebiak, 1989; Latif and Graham et al., 1992; Meinen and McPhaden, 2000; Kessler, 2002; Trenberth et al., 2002; McPhaden, 2003; Yu and Kao, 2007). The equatorial Pacific HC also is a useful predictor of Indian summer monsoon rainfall (Rajeevan and McPhaden, 2004).

There is a large body of literature studying the HC variability and its link to ENSO in the past decades (e.g. review papers by Latif et al., 1998 and Neelin et al., 1998; McPhaden, 2003; White, 1995; Lohmann and Latif, 2005). However all of these studies only cover time periods of 20–50 years due to a lack of long-term subsurface observations. The period available for studying HC variability at interannual time scale, in particular at the decadal time scale, probably precludes statistically robust conclusions. Therefore it is of interest and practical importance to explore the possibility to obtain long-term HC data, thereby effectively studying HC variability. An effective method towards this goal is to generate a long-term analysis dataset of HC using the state-of-the-art models and using other long-term observations available. With the development of assimilation technique in recent years, the reanalysis strategy has accepted intensive attention. In fact, the reanalysis NCEP and ECMWF wind datasets have been widely used as the "observations" since they were generated.

This study attempts to produce a long-term HC analysis dataset over 100 years through an OGCM, and then to further explore the variability of the upper oceanic heat content. Recently we reconstructed the surface wind stress of the tropical Pacific for the period from 1875–1947 using statistical technique and using the historic SST and sea level pressure datasets (Deng and Tang, 2008), which enables it available to implement a long-term control run of the OGCM. The reconstructed wind stress has been successfully applied to perform retrospective ENSO prediction for the past 120 years (Deng and Tang, 2008; Tang et al., 2008), suggesting the high quality and good credits of the reconstructed wind. Further, we recently also completed the assimilation of a long-term historic SST dataset into the OGCM that led to skillful retrospective predictions (Deng et al., 2008). All of these allow us to produce a long-term analysis of the upper ocean heat content for the tropical Pacific.

This paper is structured as follows: Section 2 briefly describes the model, data and assimilation scheme. Section 3 examines the quality of analyzed HC by comparing it against the NCEP reanalysis dataset for their common period from 1980–2000, as well against the observed HC from 1961–2000. In Secs. 4 and 5, the HC variability and its link to ENSO are investigated for the period from 1881–2000. A summary and discussion are given in Sec. 6.

2. Model, Data and Assimilation Scheme

2.1. *Data*

In this study we used the monthly Extended Reconstruction version2 SST (ERSST.v2) dataset from 1878–2002, reconstructed by Smith and Reynolds (2004), with a resolution of $2° \times 2°$. The data domain was configured for the tropical Pacific ocean. This bias corrected dataset has been used for studying climate variation and prediction (e.g. Xue *et al.*, 2003; Nakaegawa *et al.*, 2004; Monahan and Dai, 2004). Due to relatively poor quality of the dataset prior to 1881, we focus on the period from 1881–2001 in this study. For the validation of SST assimilation, the NCEP reanalysis subsurface temperature from 1980–2000 is also used in this study (Behringer *et al.*, 1998; referred to as the NCEP data hereafter). The data domain is confined in the tropical Pacific Ocean, spanning from 1980 to 2000 with the resolution of $1.0°$ lat. by $1.5°$ lon.

The monthly 400 m depth-averaged heat storage anomalies from the Joint Environmental Data Analysis Center at the Scripps Institution of Oceanography. This data set consists of all available XBT, CTD, MBT and hydrographic observations, optimally interpolated by White (1995) to a three-dimensional grid of $2°$ lat. by $5°$ lon., and 11 standard depth levels between the surface and 400 m. This dataset is referred to as the observation although it is still, strictly saying, a kind of reanalysis dataset.

To perform a long-term analysis with an OGCM, the past wind stress data, as the model forcing, is required. Using SST as a predictor and SVD technique, a long-term wind stress dataset from 1881–1947 was reconstructed, with the resolution $2° \times 2°$ and the domain of tropical Pacific from 30°S–30°N (see http://web.unbc.ca/ytang/wind.html). The cross-validation scheme was used in the reconstruction to ensure the training data not used in test periods. The training data of the wind is the NCEP reanalysis 10 m wind speed in monthly T62 Gaussian grids for 1948–2006 (Kalnay *et al.*, 1996). For consistency, we also used the reconstructed surface wind for the period from 1948–2001 instead of the observed wind in this study. The reconstructed wind has been applied to study ENSO predictability as mentioned in the introduction.

HC is defined here as the integral of the temperatures over the first 17 model levels, equivalent to the depth of 250 m, calculated from

$$HC = \frac{\sum_{i=1}^{17} h_i T_i}{\sum_{i=1}^{17} h(i)}, \tag{1}$$

where T_i and h_i are the temperature and depth of level i.

2.2. Model

The ocean model used in this study is the latest version of NEMO (Nucleus for European Modeling of the Ocean), identical to that used in Tang et al. (2008) and Deng and Tang, (2008). Details of the ocean model are described in http://www.lodyc.jussieu.fr/NEMO/. The domain of the model used here is configured for the tropical Pacific Ocean from 30°N–30°S and 122°E–70°W, with horizontal resolution 2.0° in the zonal direction and 0.5° within 5° of the equator, smoothly increasing to 2.0° at 30°N and 30°S in the meridional direction. There are 31 vertical levels with 17 concentrated in the top 250 m of the ocean. The time step of integration is 1.5 hours and all boundaries are closed, with no-slip conditions.

The model was first spun up for 500 years using climatological wind stress derived from the 50-year NCEP Reanalysis wind stress and the heat flux Q_s is parametrized by model temperature as follows:

$$Q_s = Q_0 + \lambda(T - T_0), \tag{2}$$

where Q_0 is the climatological heat flux, obtained from the European Center for Medium-Range Weather Forecasts (ECMWF) reanalysis project for the base period 1971–2000. T is the model SST, T_0 is Levitus observed climatological SST (Levitus 1998), and λ is the relaxation rate, set to $-40 \, \text{Wm}^2 \, \text{K}^{-1}$ (Tang et al., 2004; Moore et al., 2006). For a 50 m mixed-layer depth, this value corresponds to a relaxation time scale of two months (Madec et al., 1998).

2.3. Assimilation scheme

The ERSST.v2 data from 1881–2000 was assimilated into the OGCM using Ensemble Kalman Filter (EnKF). The assimilation domain covers the tropical Pacific from 140°E to 80°W and 15°S to 15°N in horizontal and in the upper 17 levels (250 m). The assimilation was performed by every 5 days.

Usually SST is a prognostic variable in ocean models, and the general procedure of SST assimilation is to optimally insert it into the models. However this strategy is unable to effectively correct the subsurface temperature, leading to serious imbalances between the surface and subsurface during the assimilation cycle (Tang and Kleeman, 2002). Therefore, a key issue for SST assimilation is an effective vertical transfer of information from the surface to the subsurface. Towards this goal, a special strategy of initial perturbation is used in this study to generate ensemble, namely that the perturbation field is designed to be of not only horizontal coherence but vertical coherence between adjacent levels. The vertical coherence is considered using the below method

$$\epsilon_k = \alpha \epsilon_{k-1} + \sqrt{1 - \alpha^2} W_k, \tag{3}$$

where ϵ_k is the pseudorandom field at the kth level ($k = 1, 2, \ldots, 17$), and W_k is the pseudorandom field at the kth level without considering vertical coherence, constructed using the method of Evensen (2003).

With such a well-designed perturbation scheme, the forecast error covariance matrix in EnKF can act as a time-variant transfer operator to project the SST corrections onto the subsurface temperatures effectively (Deng *et al.*, 2008). As such the increments of subsurface temperatures can be obtained via the transfer operator during assimilation cycles. The details of the assimilation system by EnKF can be found in Deng *et al.* (2008).

3. Validation of the HC Analysis from 1980–2000

A long-term control run was performed with the OGCM, forced by the reconstructed wind from 1881–2000. In this control run, the SST was also assimilated into the OGCM sequentially every five days from 1881–2000 using EnKF with a well-designed perturbation scheme as aforementioned. From the long-term assimilation run, we obtained a long term analysis for all variables of the OGCM including SST and subsurface temperatures. The SST analysis and the retrospective ENSO predictions initialized from these analyses were discussed in details in Deng *et al.* (2008). In this section, we will validate HC analysis through comparing it with the NCEP HC dataset of the upper 250 m that assimilated both altermetry data and observed

92   Y. Tang and Z. Deng

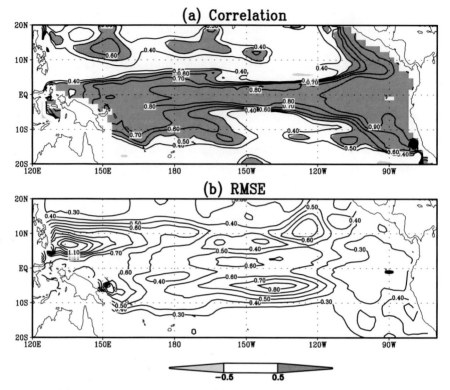

Fig. 1. (a) Anomaly correlation R and (b) $RMSE$ between the analyzed HC and the NCEP HC for the period from 1980–2000.

SST, and with the monthly 400 m depth-averaged heat storage anomalies from Scripps Institution that consists of all available XBT, CTD, MBT and hydrographic observations.

Figure 1 shows the correlation and RMSE (root mean square error) between the analyzed HCA and the NCEP counterpart for the period from 1980–2000. The best analysis skill appears in the eastern Pacific and the whole equatorial belt, with correlation coefficient over 0.7. The HC analysis is relatively poorer in the region 10° off the equator. This is due mainly to two reasons: (i) the assimilation domain is confined within the equatorial belt of 15° based on the consideration of computation expense; (ii) the OGCM only has good simulation skills in the equatorial belt. The poor simulated skills off the equator are common defects in almost all ocean models including OGCMs (e.g. Deng et al., 2008; Lou et al., 2005). The good HC analysis in the equatorial Pacific is important and has practical

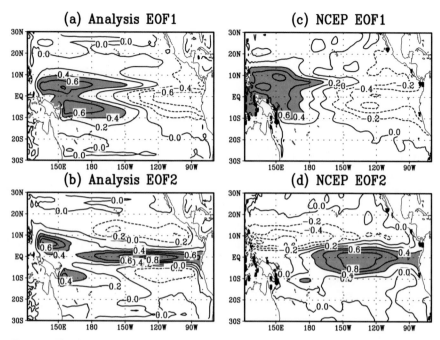

Fig. 2.   The first and second EOF modes for the analyzed HC (a and b) and the NCEP HC (c and d).

significance since the strongest coupling of the air-sea occurs there, and the HCA distribution along the equator dominates ENSO characteristics and evolution.

Figure 2 compares two leading EOF modes of the analyzed HCA with those derived from the NCEP HCA during 1980–2000. The two EOFs of the analyzed HCA account for 78% and 14% of the total variance, respectively, which compare with 33% and 15% of the total variance accounted for by two leading modes from the NCEP HCA. The larger variance accounted for by the leading modes of the analyzed HCA is probably because the reconstructed wind stress that removed the high frequent components forced the ocean model. As seen in Fig. 2, the leading EOFs for the analyzed HCA (Figs. 2(a) and 2(b)) generally resembled the NCEP modes (Figs. 2(c) and 2(d)), except a stronger zonal HCA gradient at the equatorial central Pacific. The major characteristics of the first mode (Figs. 2(a) and 2(c)) has a dipole zonal structure involving a western Pacific "Rossby wave-like" response of one sign and an eastern Pacific "Kelvin wave-like" response of the opposite sign; whereas the second mode (Figs. 2(b) and 2(d)) has a large-amplitude signal of the one sign located mainly in the equatorial central/eastern Pacific. These patterns agree with the idea of a heat content buildup prior to El Niño as postulated by Wyrtki (1985) and Jin (1997), and are consistent with the delayed-action oscillator mechanism (e.g. Battisti, 1988; Suarez and Schopf, 1988). They are also very similar to those reported in previous work (e.g. Tang *et al.*, 2005). Comparing leading modes between the analyzed HCA and the NCEP HCA reveals that the former has a stronger zonal HCA gradient at the equatorial central Pacific and a more obvious wave-like structure along the equator, probably because it was derived from the OGCM forced with the high-frequency free reconstructed wind stress.

Comparison of the leading principal components (PCs) between the analyzed and NCEP HCA is shown in Fig. 3. As can be seen, the analyzed HCA PCs are in very good agreement with the PCs of NCEP HCA, with their correlation coefficients over 0.8. This is also true for the averaged HCA over several Niño regions as shown in Fig. 4. However the analyzed HCA

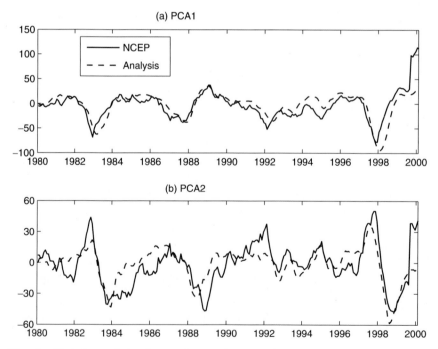

Fig. 3.   Variations of the first and second principal components for the analyzed and the NCEP HC.

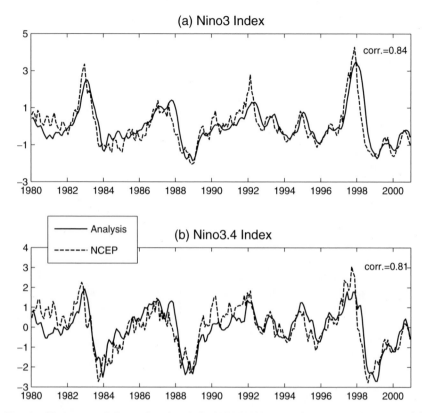

Fig. 4.   Variations of the analyzed and the NCEP HC anomalies, averaged over the (a) Niño3 and (b) Niño3.4, for the period from 1980–2000.

leads NCEP HCA by 1–2 months, which may be due to the model bias in the OGCM. It was found that the model bias often results in the simulation of temperature anomaly variation leading the observation by 1–2 months in many oceanic models (e.g. Tang *et al.*, 2001).

Figure 5 shows the time-longitude plot of HCA along the equator during 1980–2000, taken from the analyzed and the NCEP data. As shown in Fig. 5, the analyzed HCA agreed well with the NCEP HCA, and captured all observed ENSO signals. The common deficiency for the analysis is relatively weak simulated amplitude, which is a common problem in many OGCMs.

We repeated all validations performed above using the observed heat storage (HS) of the upper 400 m for 1961–2000. The results are similar to those shown above. For example, Fig. 6 shows the correlation coefficients between the analyzed HCA against the observed HS anomalies, which is

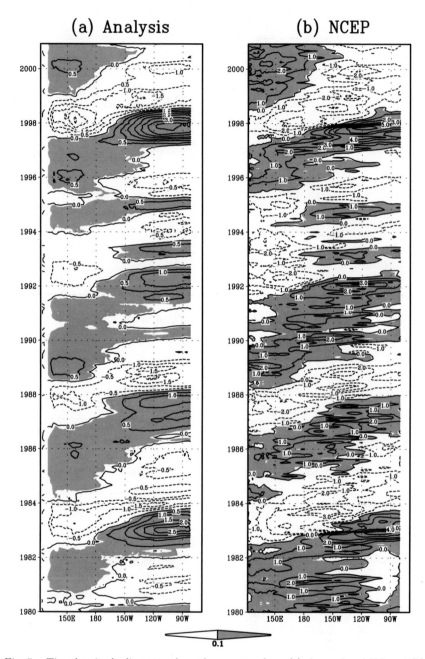

Fig. 5.   Time-longitude diagrams along the equator, from (a) the analyzed HC and (b)
the NCEP HC. Contour interval is 0.5° C in (a) and 1.0° C in (b). The positive anomalies
above 0.1° C are shaded.

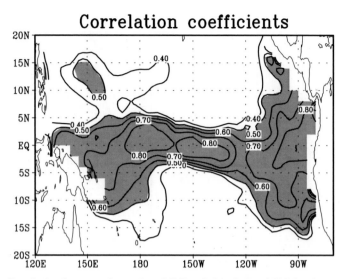

Fig. 6. Correlation between the analyzed HC and the observed HC for the period from 1961–2000.

very similar to Fig. 1(a). It should be noticed that HS data is defined as the integral of temperature over the depth multiplied by a constant coefficient and has a different unit (Watt-Seconds/Meter^2) from the analyzed and the NCEP HC, thus it is meaningless to compare their amplitude.

In summary, the analyzed HC can well characterize the realistic variations of the upper oceanic heat at monthly or longer time scales. It allows us to explore the upper ocean thermal states, in particular the HC variability at the interannual or longer time scales.

4. Variability of HC from 1881–2000

Figure 7 shows the evolution of the analyzed HCA along the equator from 1881–2000. The most striking feature in this time-longitude diagram is the interannual variability of HCA with a period of 2–5 years during the whole period. Comparing the interannual variability of HCA with ENSO variability of SSTA (not shown) reveals a very good relationship between them, namely that, the analyzed HCA captures well all ENSO events from 1881–2000. A further scrutiny of Fig. 7 finds that the interannual variability of HCA has decadal/interdecadal variation as in that of SSTA (Tang *et al.*, 2008). For example, the magnitude of HCA is visibly larger during the period from 1980–2000 than during other periods whereas the

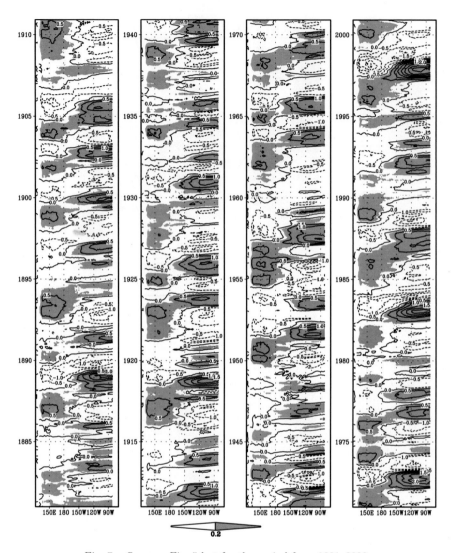

Fig. 7.   Same as Fig. 5 but for the period from 1881–2000.

HCA is likely to have the smallest magnitude in the 1920s and 1930s. Such a decadal/interdecadal variation in HCA interannual variability is more obvious in Fig. 8, which shows the wavelet power spectrums of the first two principal components of HCA. The local wavelet power spectrum clearly indicates that the significant periods are localized in time. During 1960–2000, the signal is significant at the periods of 2–5 years whereas

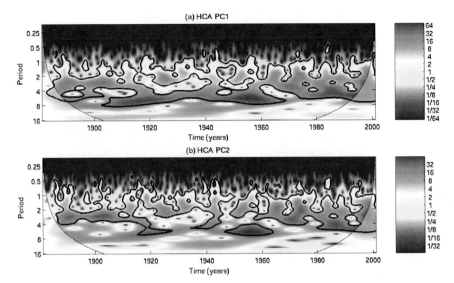

Fig. 8. The wavelet power spectrum of (a) HCA PC1 and (b) HCA PC2. The power spectrum is normalized by 95% confidence critical power calculated by Monte Carlo significant test method. The area under the arc line is the cone of influence, where zero padding has reduced the variance. Black contour is the 5% significance level, using a red-noise (autoregressive lag 1) background spectrum. The period of unit (y-axis) is year.

during 1905–1960 the strong signal appears at the periods of 4–8 year with weak signal at the periods of 2–4 years. One might speculate that the interdecadal variation of HCA interannual variability is due probably to the data quality since the observations were very sparse and sporadic, even unavailable before 1960. However, there were strong ENSO signals at the periods 2–5 years before 1905, which might effectively remove such a speculation.

The interdecadal variations in HCA interannual variability are further displayed in Fig. 9, which shows the variation of the strength of ENSO signal measured in each running window of 20-yr from 1881–2000 (i.e. 1881–1900, 1882–1901,..., 1981–2000).[1] For comparison, the ENSO signal of SSTA is also presented. Two methods were applied to extract the ENSO signal in this study. The first was to perform spectrum analysis for the first principal component (PC1) of the analyzed HCA and observed SSTA, respectively,

_____

[1]The signal measured during a 20-yr window is plotted at the middle point of the window in Fig. 9. For example, the signal at 1890 was calculated using the samples from 1881–1900. The 20-yr window is shifted by one year each time starting from 1881 until 2000.

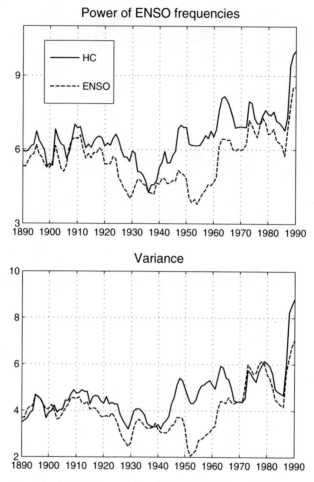

Fig. 9.   The strength of the interannual variability of SSTA and HCA, measured by (a) the spectrum power at ENSO frequencies of 2–5 years of PC1 and (b) the variance of PC1, both calculated in each 20-yr running window from 1881–2000.

for each 20-yr running window, using the total spectrum power at the frequencies of 2–5 years to represent the strength of ENSO signal, as shown in Fig. 9(a). The second was to use the variance of the analyzed HCA PC1 and observed SSTA PC1, computed for each running window of 20-yr from 1881–2000, as shown in Fig. 9(b). The two methods produce very consistent results. Figure 9 shows a significant interdecadal variation in ENSO signal in both the surface temperature and the subsurface heat content, and a consistent relationship between the variation in SSTA and that in HCA.

In the late 19th century and the early 20th century, ENSO signal was relatively strong and stable. Since the early 1920s, the signal strength was weakened, reaching a minimum around 1940s, beyond which the signal rebounded and increased with time until the 1960s. ENSO signal was the strongest from the 1960s, especially in the late 20th century. Therefore, there is a striking interdecadal variation of ENSO signal in the upper heat content anomalies of the tropical Pacific during the past 120 years from 1881–2000. In the next section, we will see the interdecadal variation of ENSO signal is a major reason to cause the interdecadal variation in ENSO prediction skill.

5. HC Variability and ENSO Predictability

It has been reported in several recent works that ENSO predictability has decadal/interdecadal variation, which was argued to be due mainly to the corresponding decadal/interdecadal variation in ENSO variability (e.g. Chen *et al.*, 2004; Tang *et al.*, 2008; Deng and Tang, 2008). In this section, we will examine the relationship of decadal/interdecadal variations between ENSO predictability and HCA variability, which has been so far little addressed. Towards this goal, a long-term retrospective ENSO prediction was performed for the period from 1881–2000 using the hybrid coupled model (HCM), i.e. the OGCM coupled with a statistical atmospheric model. The statistical atmospheric model is a linear model that predicts the contemporaneous surface wind stress anomalies from SSTA, which was constructed by the singular vector decomposition (SVD) method with cross-validation scheme. During the initialization of the hybrid coupled model, the OGCM was forced by the sum of the associated wind anomalies computed from the atmospheric model and the observed monthly mean climatological wind stress. Full details of the HCM are given in Deng *et al.* (2008) and Tang *et al.* (2004, 2008).

A total of 480 forecasts, initialized from January 1881 to October 2000, were run starting at three months intervals (1 January, 1 April, 1 July, 1 October), and continued for 12 months for the HCM. The SST assimilation was performed to initialize the forecasts as introduced in Sec. 2.3.

Figure 10 shows the averaged correlation R and *MSE* (mean square error) over 1–12 months lead measured in each running window of 20-yr from 1881–2000 (thin blue line and dashed green line), i.e. 1881–1900, ..., 1981–2000, where the predicted Niño3.4 SSTA (5°N–5°S,

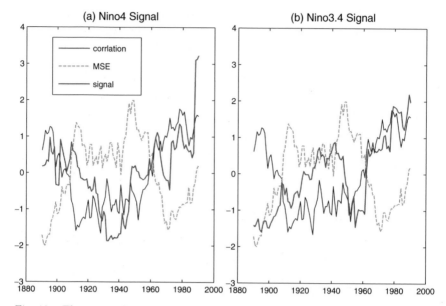

Fig. 10.   The averaged prediction skill of Niño3.4 SSTA index over the first 12 month
leads, against the HC signal measured by (a) HCA Niño4 index and (b) HCA Niño3.4
index. The evaluation was done in each 20-yr running window from 1881–2000. The
normalization was applied prior to plotting for removing the unit.

170°W–120°W) is compared against the observed value. As can be seen,
there is a striking interdecadal variation of ENSO predictability in the
past 120 years from 1881–2000 in the HCM. Generally there is a high
predictability in the late 19th century and in the middle-late 20th century,
and a low predictability from 1901–1960. Figure 10 also displays the signal
of HCA in Niño4 (5°N–5°S, 160°E–150°W) (Fig. 10(a)) and in Niño3.4
(Fig. 10(b)) (thick red line), respectively, measured by their individual
variance.

    Figure 10(a) demonstrates a significant relationship between the ENSO
predictability and the signal of Niño4 HCA. Both display a consistent
interdecadal variation. In the late 19th century, the signal was strong,
and the model showed a large correlation R and a low MSE. Since
then the signal strength weakened and the skill continuously declined
with time, both reaching a minimum around 1940s, beyond which both
rebounded and increased with time until the 1960s. The model has a
relatively good prediction skill from the 1960s, especially in the late 20th
century. Correspondingly, the Niño4 HCA signal is also the strongest

in these periods. Such a good relationship between Niño4 HCA signal and prediction skills holds not only for correlation R but also for *MSE* skill.

Figure 10(b) shows the Niño3.4 HCA signal (thick red line), which is somehow different from that of Niño4. For example, the Niño3.4 HCA signal was week in the late 19th century but relatively strong during the period from 1901–1940, which was almost out of the phase of that of Niño4. The overall relationship between Niño3.4 HCA signal and ENSO predictability is weak in Fig. 10(b), although there is a good relationship after the late 1960s. This is different from the relationship between Niño3.4 SSTA signal and ENSO predictability, which is significantly strong for the whole period (Tang *et al.*, 2008). We also explored the relationship between ENSO predictability and the HCA signal measured using PC1 and Nino3 index, respectively, and got similar results.

It is of interest to explore the underlying physical interpretation of the relationship between Niño4 HCA signal and prediction skill. As discussed in the introduction, the most important physical and dynamical process responsible for ENSO cycle probably is the "Discharge" mechanism of the upper HC of the equatorial western Pacific or "Delayed oscillator mechanism", both asking a significant lagged relationship between the Niño4 HCA signal and the Niño3 (Niño3.4) SSTA signal. Indeed, it has been found in many ENSO prediction models that the model predication skill is usually high when the lagged correlation is strong, and the Niño4 HCA is a very good precursor of Nino3 (3.4) SSTA evolution (e.g. Latif *et al.*, 1998; Tang and Hsieh, 2003).

Shown in Fig. 11(b) is Niño3.4 SSTA prediction skill measured in 6 sub-periods of 20 years each. It is evident in Fig. 11(b) that the correlation skills are significantly different among these periods. Comparing Figs. 10 and 11(b) reveal a considerable consistency of variations in correlation skill. For example, high prediction skills appear in the late 19th century and the middle-late 20th century, i.e. 1881–1900, 1961–1980 and 1981–2000, whereas the periods of 1901–1920, 1921–1940, and 1941–1960 have relatively low prediction skills. A similar consistency is also found in the *MSE* skill (not shown). Figure 11(a) shows the lagged correlation between the analyzed Niño4 HCA and the observed Niño3 SSTA, with SSTA leading to HCA. As can be seen, the period that has a high prediction skill has also a good lagged correlation and vice versa. For example, the best correlation prediction skills appear in the middle-late 20th century when the lagged correlation is the highest during this period whereas the periods that have

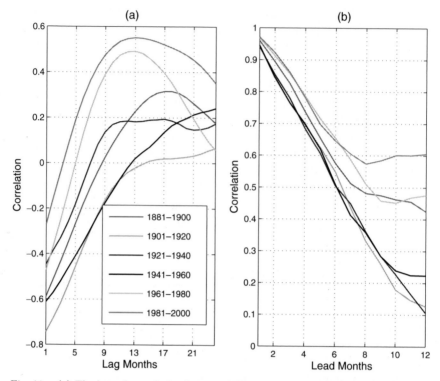

Fig. 11. (a) The lagged correlation between Niño4 HCA index and Niño3 SSTA index, with the HCA behind SSTA, for six different periods; (b) the correlation skill between predicted Niño3.4 SSTA index against the observed value, as a function of leading time, for the corresponding periods.

the minimum lagged correlation have very poor skills such as 1941–1960 and 1901–1921.

In summary, there is a striking interdecadal variation of ENSO predictability in the past 120 years from 1881–2000, which is highly related to interdecadal variation of the signal of Niño4 HCA. When the signal of Niño4 HCA is stronger, the lagged correlation between it and the equatorial eastern Pacific SSTA is larger, leading to better prediction skills.

6. Discussion and Summary

An important step in understanding ENSO and the interaction of air-sea of the tropical Pacific ocean is to analyze the upper ocean heat content, as evidenced in a large body of literature. However all studies about the tropical pacific upper ocean HC have been so far confined within the last 20–50 years due to the unavailability of the longer data, which is not long enough to study HC variability at the interannual and decadal scales.

In this study, we explored the possibility of producing a long-term HC analysis dataset over 100 years through an OGCM and a well-designed EnKF-based assimilation system for the historic SST dataset. The results show that the analyzed HC, compared with the NCEP data and the observation, well characterizes the realistic variability of the HC at monthly or longer time scales. The correlation coefficients between the analyzed HC with the NCEP HC are very high up to 0.9 in the equatorial eastern Pacific ocean. It is also true when the analyzed HC is validated against the observed HC.

Further we examined the variation of the tropical Pacific upper ocean HC from 1881–2000. It was found that there exists a striking interannual variability in the tropical Pacific upper ocean HCA. Like ENSO variability, the HCA interannual variability has also a significant interdecadal variation. In the late 19th century and the early 20th century, the HC interannual signal was relatively strong and stable. Since the early 1920s, the signal strength was weakened, reaching a minimum around 1940s, beyond which the signal rebounded and increased with time until the 1960s. The interannual variability was the strongest from the 1960s, especially in the late 20th century.

We also analyzed a set of long-term retrospective forecasts of the past 120 years with the HCM. It was found that the model prediction skill displays a consistent interdecadal variation with that of HCA variability, namely that the prediction skill was high in the late 19th century from 1881–1900, and then declined with time, reaching a minimum around 1940–1950s, beyond which it rebounded and increased with time until the 1960s. It had relatively high prediction skill from the 1960s, especially in the late 20th century from 1981–2000. A good relationship between ENSO predictability and the lagged correlation of Niño4 HCA-Niño3 SSTA was also found. These indicate that the interdecadal variation in predictability is highly related to the interdecadal variation of HCA variability itself. A strong HCA signal in the equatorial western Pacific produced a large lagged correlation of Niño4 HCA-Niño3 SSTA, leading to more precursory information at the initial time of predictions. As such, the prediction is likely to be more reliable.

Several cautions should be borne in mind when using the long term HC analysis. First, the oceanic analysis was obtained through forcing the reconstructed wind and historic SST data where the former was constructed by the line statistical method. The SST data is coarse and gappy before 1950s, thus it might only contain useful information on large-scale interannual or decadal/interdecadal climate variability. Such information might be sufficient to describe and characterize some large-scale climate modes such as ENSO, but not enough for relatively short and small scale variability such as some tropical oceanic waves. Thus the HC analysis should be mainly used for studying large-scale climate signals. Second, the long-term trend was removed from SST prior to constructing the wind in order to manifest the signal of the interannual variability, thereby precluding the long-term trend in the HC analysis. This suggests that the HC analysis might not be suitable for studying the issues related to the global warming that has been detected in the upper ocean (e.g. White et al., 2003). Third we used a running window of 20-yr to analyze interdecadal variations in predictability and the HC variability. The window length of 20 years was motivated by Chen et al. (2004) where the interdecadal variations in predictability were discussed in such interval. We also performed several sensitivity experiments, with the window length of 10-yr, 30-yr and 40-yr. The relationships of predictability to the HCA signal are similar to those presented in this paper. Finally the reconstructed winds, subject to a common statistical problem, generally underestimate the amplitude of wind anomaly, thus underestimating the amplitude of HCA. Nevertheless, this study is to date the first work to produce the tropical Pacific upper ocean heat content analysis for the past 120 years. The HC analysis has led to some interesting findings about ENSO variability and predictability as presented in this paper. It has been posted on the internet and is freely loadable (http://web.unbc.ca/ytang/wind.html). Therefore, this work has both theoretical contribution and practical significance in studying the tropical Pacific climate variability, especially for ENSO.

Acknowledgment

This work is supported by BC — China Innovation and Commercialization Strategic Development Program.

References

1. D. S. Battisti, Dynamics and thermodynamics of a warming event in a coupled tropical atmosphere-ocean model, J. Atmos. Sci. **45** (1988) 2889–2919.
2. D. W. Behringer, M. Ji and A. Leetmaa, An improved coupled model for ENSO prediction and implications for ocean initialization. Part I: The ocean data assimilation system, Mon. Weather Rev. **126** (1998) 1013–1021.

3. D. Chen, M. A. Cane, A. Kaplan, S. E. Zebiak and D. Huang, Predictability of El Nino in the past 148 years, *Nature*, vol. 428, pp. 733–736.

4. Z. Deng, Y. Tang and X. Zhou, The retrospective prediction of ENSO from 1881–2000 by a hybrid coupled model (I): SST assimilation with Ensemble Kalman Filter, *Climate Dyn.* (2008), doi 10.1007/s00382-008-0399-1.

5. Z. Deng and Y. Tang, The retrospective prediction of ENSO from 1881–2000 by a hybrid coupled model (II): Interdecadal and decadal variations in predictability, *Climate Dyn.* (2008), doi 10.1007/s00382-008-0398-2.

6. G. Evensen, The ensemble Kalman filter: Theoretical formulation and practical implementation, *Ocean Dyn.* **53** (2003) 343–367.

7. F. F. Jin, An equatorial recharge paradigm for ENSO. Part I: Conceptual model, *J. Atmos. Sci.* **54** (1997) 811–829.

8. E. Kalnay *et al.*, The NCEP/NCAR 40-year reanalysis project, *Bull. Am. Meteor. Soc.* **77** (1996) 437–470.

9. W. S. Kessler, Is ENSO a cycle or a series of events? *Geophys. Res. Lett.* **29** (2002) 2125, doi:10.1029/2002GL015924.

10. M. Latif, D. Anderson, T. Barnett, M. Cane, R. Kleeman, A Leetmaa, J. O'Brien, A Rosati and E. Schneider, A review of the predictability and prediction of ENSO, *J. Geophys. Res.* **103** (1998) 14375–14393.

11. M. Latif and N. E. Graham, How much predictive skills is contained in the thermal structure of an Oceanic GCM? *J. Phys. Oceanogr.* **22** (1992) 951–962.

12. S. Levitus and T. Boyer, NOAA/OAR/ESRL PSD, Boulder, Colorado, USA (1998), http://www.cdc.noaa.gov.

13. K. Lohmann and M. Latif, Tropical Pacific decadal variability and the subtropical-tropical cells, *J. Climate* **18** (2005) 5163–5178.

14. J.-J. Luo, S. Masson, S. Behera, S. Shingu and T. Yamagata, Seasonal climate predictability in a coupled OAGCM using a different approach for ensemble forecasts, *J. Climate* **18** (2005) 4474–4495.

15. G. Madec, P. Delecluse, M. Imbard and C. Levy, OPA 8.1 Ocean General circulation model reference manual, Institut Pierre Simon Laplace (IPSL) (1998), p. 91.

16. M. J. McPhaden, Tropical pacific ocean heat content variations and ENSO persistence barriers, *Geophys. Res. Lett.* **30** (2003) 1480, doi:10.1029/2003GL016872.

17. C. S. Meinen and M. J. McPhaden, Observations of warm water volume changes in the equatorial Pacific and their relationship to El Niño and La Niña, *J. Climate* **13** (2000) 3551–3559.

18. A. H. Monahan and A. Dai, The spatial and temporal structure of ENSO nonlinearity, *J. Climate* **17** (2004) 3026–3036.

19. A, J. Moore, Y. Zavala-Garay, R. Tang, J. Kleeman, A. Vialard, K. Weaver, D. L. Sahami, T. Anderson and M. Fisher, Optimal forcing patterns for coupled models of ENSO, *J. Climate* **19** (2006) 4683–4699.

20. T. Nakaegawa, M. Kanamttsu and T. M. Smith, Interdecadal trend of prediction skill in an ensemble AMIP-type experiment, *J. Climate* **15** (2004) 2881–2889.

21. J. D. Neelin, D. S. Battisti, A. C. Hirst, F.-F. Jin, Y. Wakata, T. Yamagata and S. Zebiak, ENSO theory, *J. Geophys. Res.* **103** (1998) 14261–14287.

22. M. Rajeevan and M. J. McPhaden, Tropical pacific upper ocean heat content variations and Indian summer monsoon rainfall, *Geophys. Res. Lett.* **31** (2004) L18203, doi:10.1029/2004GL020631.

23. T. M. Smith and R. W. Reynolds, Improved extended reconstruction of SST (1854–1997), *J. Climate* **17** (2004) 2466–2477.

24. M. J. Suarez and P. S. Schopf, A delayed action oscillator for ENSO, *J. Atmos. Sci.* **45** (1988) 3283–3287.

25. Y. Tang and R. Kleeman, A new strategy for SST assimilation for ENSO prediction, *Geophy. Res. Lett.* (2002), 10.1029/2002GL014860.

26. Y. Tang and W. W. Hsieh, ENSO simulation and predictions using a hybrid coupled model with data assimilation, *J. Japan Meteorol. Soc.* **81** (2003) 1–19.

27. Y. Tang, R. Kleeman and A. Moore, SST assimilation experiments in a tropical Pacific Ocean model, *J Phys. Oceangr.* **34** (2004) 623–642.

28. Y. Tang, R. Kleeman and A. Moore, On the reliability of ENSO dynamical predictions, *J. Atmos. Sci.* **62** (2005) 1770–1791.

29. Y. Tang, Z. Deng, X. Zhou, Y. Cheng and D. Chen, Interdecadal variation of ENSO predictability in multiple models, *J. Climate* (2008).

30. K. E. Trenberth, J. M. Caron, D. P. Stepaniak and S. Worley, Evolution of El Nino southern oscillation and global atmospheric surface temperatures, *J. Geophys. Res.* **107** (2002) 4065, doi:10.1029/2000D000298.

31. W. B. White, Design of a global observing system for gyre-scale upper ocean temperature variability, *Prog. Oceanogr.* **36** (1995) 169–217.

32. W. B. White, M. D. Dettinger and D. R. Cayan, Sources of global warming in the upper ocean on decadal period scales, *J. Geophys. Res.* **108** (2003) 3248, doi: 10.1029/2002JC001396.

33. K. Wyrtki, Water displacements in the Pacific and the genesis of El Niño cycles, *J. Geophys. Res.* **90** (1985) 7129–7132.

34. Y. Xue, T. M. Smith and R. W. Reynolds, Interdecadal changes of 30-yr SST normals during 1871–2000, *J. Climate* **15** (2003) 1601–1612.

35. J.-Y. Yu and H.-Y. Kao, Decadal changes of ENSO persistence barrier in SST and ocean heat content indices: 1958–2001, *J. Geophys. Res.* **112** (2007) D13106, doi:10.1029/2006JD007654.

36. S. E. Zebiak, Oceanic heat content variability and El Niño cycles, *J. Phys. Oceanogr.* **19** (1989) 475–486.

Advances in Geosciences
Vol. 18: Ocean Science (2008)
Eds. Jianping Gan et al.
© World Scientific Publishing Company

SALINITY VARIATIONS IN WATER COLUMN DUE TO OUTFLOWS ESTIMATED BY MULTI-SENSOR REMOTE SENSING*

XIAO-HAI YAN[†] and YOUNG-HEON JO[†,‡]
*Center for Remote Sensing, Department of Oceanography,
College of Marine and Earth Studies, University of Delaware,
Newark, DE 19716, U.S.A*
‡joyovng@udel.edu

W. TIMOTHY LIU[†]
*4800 Oak Grove, Dr. Jet Propulsion Laboratory
Pasadena, CA 91109, U.S.A.*

MINHAN DAI[†]
*State Key Laboratory of Marine Environmental Science,
Xiamen University, Xiamen, 361005, China*

Since there are no direct salinity measurements using remote sensing from space, we developed a new method to estimate salinity variations in the water column using satellite multi-sensor measurements. The technique for estimating vertical salinity variations (and thus salt steric height) is derived from sea surface height variation measured by satellite altimetry after removing the thermal and other steric components. We call this technique Integrated Multi-Sensor Data Analysis (IMSDA). We apply the IMSDA to estimate salinity variation from the Amazon River outflow in the tropical Atlantic, the Yangtze River outflows in the East China Sea (ECS), and Mediterranean Outflow and Meddies (O&M) in the North Atlantic.

*This research was supported partially by NASA through its Physical Oceanography Program, EPSCoR Program and Space Grant Program, by NOAA Sea Grant, and by the Xiamen University 985, and 973 Programs.
†Also affiliated with the Joint Institute for Coastal Research and Management (Joint-CRM) of University of Delaware, Newark, Delaware and Xiamen University, Xiamen, China.

For the Amazon River outflow, we estimated salt steric height using IMSDA and compared it with chlorophyll concentration (Chl_a) observed by the Sea-viewing Wide Field-of-view Sensor (SeaWiFS). In addition to ocean color observations, we compared salt steric height with the Sea Surface Salinity (SSS) measured at the Prediction and Research Moored Array in the Atlantic (PIRATA) mooring stations. Comparisons of long-term time series of salt steric height and Chl_a were highly correlated with mooring data at 8°N 38°W.

For the Yangtze River outflow, we discussed salt steric height in the ECS before and after the Three Gorges Dam Water Storage (TGDWS). The patterns of the salt steric height anomaly near the Yangtze River Estuary (YRE) agreed well with the *in situ* SSS measurements. In order to examine the changes in stability in the mixed layer near the YRE due to TGDWS, the time series of the steric height anomaly ratio (R'_s) between heat and salt steric height near the Yangtze River mouth were estimated. The R'_s was strongly associated with a coastal upwelling in the ECS, because the reduced freshwater outflow gives rise to changes of both mixed layer depth and surface current patterns off the YRE.

For the Mediterranean Outflow and Meddies, our method successfully estimates the outflow transport through the Gibraltar Strait and Meddies in the North Atlantic Ocean. We discussed how we can detect Meddies, which are difficult to detect with the poor spatial and temporal resolution of conventional *in-situ* observations. Our estimation of outflow from the Mediterranean Sea showed that while more salty Mediterranean water was transported into the north Atlantic in late fall, the Atlantic water was transported into the Mediterranean Sea in early spring over the Strait of Gibraltar.

The method we developed to estimate salinity variation may lead to a better understanding of the global ocean circulation and global climate change.

1. Introduction

Sea Surface Salinity (SSS) is a key parameter in estimating the influence of oceans on climate [34]. Unlike temperature, salinity has no direct effect on air/sea exchanges, but it determines the convection and recurrence of water masses, which significantly effect seasonal and interannual variability of the global system. Thus, three principal scientific objectives related to or affected by ocean salinity have been identified [19]: (1) Improving seasonal to interannual climate predictions; (2) Improving ocean rainfall estimates and global hydrologic budgets; and (3) monitoring large-scale salinity events.

Sea surface salinity measurements are expensive and very heterogeneously distributed, so the spatial distribution and the time variability of SSS are still very poorly known over most of the ocean surface. In fact, *in-situ* observations of SSS remain very sparse [20] and although satellite remote sensing offers the advantage of systematic global surface sampling, no satellite is yet available.

There are several algorithmic or numerical ways to circumvent this lack of salinity measurements. Several studies have been conducted in the western Pacific to construct the temperature and salinity profiles using T/P altimetry observations [24, 39, 27]. In order to obtain salinity information, Shi *et al.* [35] estimated sea salt using altimetry and *in situ* measurements. While awaiting the future missions for SSS measurements, SMOS (ESA) and Aquarius (NASA), we can still estimate this by means of altimetry and temperature measurements as demonstrated in this study.

The mean SSS distribution for the global ocean Levitus [23] is shown in Fig. 1(a), which represents data compiled from many individual ship measurements during the past 125 years. The smooth contours hide the fact that even using data collected over this long period leaves large portions of the ocean poorly sampled [20]. Satellite monitoring to provides global maps of salinity that would help resolve large-scale features of the salinity field and provide new information on its variability with time, which is not practical to obtain with *in-situ* measurements techniques. Measuring subsurface salinity is more difficult than that in sea surface. In fact, SSS features are quite different from the integration of salinity in a water column as shown in Fig. 1(b). Using Levitus climatological salinity, we integrated it from surface to 400 m depth, which is defined as salt steric height. The differences in Figs. 1(a) and 1(b) result from the presence of different water masses under sea surface. Thus, the motivation of this study is to estimate salinity variation through salt steric height in the water column using satellite remote sensing.

First, we introduced a theoretical frame work for our methodology, data and error analysis in Sec. 2. In Sec. 3, we applied IMSDA to estimate salinity variation due to the Amazon River outflow, the Yangtze River outflow, and Mediterranean outflow and Meddies.

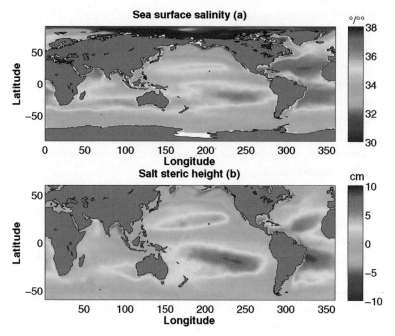

Fig. 1. (a) Sea surface salinity (SSS) from Levitus 94 climatology data, and (b) salt
steric height using Eq. (4).

2. Theoretical Background, Methodology, Data and Error Analysis

As discussed in great detail by Gill and Niiler [8], the time-varying
component of the sea level variation (η') can be written as a sum of three
terms:

$$\eta' = \frac{1}{g\rho_0}P_a' + \eta'_{ST} + \frac{1}{g\rho_0}P_b', \tag{1}$$

where P_a' is atmospheric pressure anomaly, P_b' is bottom pressure anomaly,
g is the gravitational acceleration, and ρ_0 is the density of freshwater. The
first term on the right represents the barometric component of the sea level
variability, the second term represents changes in sea level caused by steric
component (η'_{ST}) due to thermal expansion and contraction of the water
column, and the third term is due to wind action. In order to decompose
each dynamic component, the following equations were used for this study.
The barometric component is easily computed. The sea level height expands
by approximately 1 cm for 1 mbar decrease in sea level pressure. The second

term (near surface steric anomaly) is

$$\eta'_{ST} = \frac{1}{\rho_0} \left[\int_{-h}^{0} \frac{\partial \rho}{\partial T} T' dz + \int_{-h}^{0} \frac{\partial \rho}{\partial S} S' dz \right], \tag{2}$$

where T and S represent the temperature and salinity, respectively. h is the reference level, and ρ is the seawater density. The steric height anomaly is made up of two terms that arise separately from near-surface changes in heat and salinity. Equation (2) can be written individually as

$$\eta'_T = \alpha \int_{-400}^{0} T' dz, \tag{3}$$

$$\eta'_S = \beta \int_{-400}^{0} S' dz, \tag{4}$$

where α is the thermal expansion coefficient, and β is the salt expansion coefficient. Equation (3) represents thermal steric height anomaly, and Eq. (4) represents salt steric height anomaly.

The third term of Eq. (1) represents the effect of density changes below the seasonal thermocline resulting from wind action, which results in Sverdrup balance (η'_{SV}) and Ekman pumping (η'_{EK}), i.e.

$$\eta'_{SV} = \frac{1}{\beta^*} curl(\tau'), \quad \text{and} \quad \eta'_{EK} = \frac{-g'}{\rho_0 g} curl(\tau'/f). \tag{5}$$

where β^* is the meridional derivative of a Coriolis parameter, τ' is the wind stress anomaly, and g' is the reduced gravitational acceleration. On the basis of scaling arguments, Gill and Niiler [8] showed that poleward of 30°N, the seasonal ocean response was expected to be barotropic (associated with the Sverdrup term), whereas baroclinic mechanisms dominate equatorward of 30°N (Ekman pumping). In addition to wind driven sea level variation through Ekaman and Sverdrup transport, we discussed wind driven bathymetry effect on the altimetry in *error analysis* section.

What does Eq. (1) mean? The total water column is determined by the three components as discussed above. In fact, we can estimate most of these components/properties using satellite observations, with the exception of salinity. Since satellite altimeter and scatterometer provide sea surface height and wind variation, we can easily estimate total water column and wind induced sea surface height variation. Since thermal steric height variation can be determined by XBT temperature measurements, we can solve Eq. (1) for salinity variation due to freshwater discharge into the sea, so called Integrated Multi-Sensor Data Analysis (IMSDA), i.e.

$$\Delta\eta'_S = \eta'_{\text{Total}} - (\eta'_T + \eta'_W). \tag{6}$$

η'_{Total} is the deviation of the sea surface topography measured by satellite altimetry. The η'_T is defined in Eq. (3), which can be estimated from temperature measurements from satellite IR imagery and XBTs using Coupled Pattern Analysis (CPA) as discussed below. The η'_W is wind driven sea surface height anomaly variation, which can be estimated from wind scatterometer data from European Remote Sensing Satellites (ERS1/2), NASA Scatterometer (NSCAT), and Quickscat.

Coupled pattern analysis: CPA provides another means of matching patterns for two separate time-dependent fields, provided the processes underlying the two are simply and directly related. In essence, they are a means of deducing matched pairs of spatial patterns, one for each field, that have a high temporal correlation. This property makes it more attractive than deriving the EOF modes of each of the two fields separately and comparing them, mode by mode. This method is effective, though the purely statistical nature of the analysis and the empirical nature of the basis functions for the two fields mean that the EOF modes of the two fields may not be necessarily related to each other. CPA has been used in analyzing climate data [4], and matching the patterns of variability of SST and atmospheric pressure [30, 40].

CPA analysis is similar to EOF analysis [22] in that the two fields are decomposed into orthonormal modes:

$$f(x_i, t_i) = \sum_{k=1}^{M} \alpha_k(t_n)u_k(x_i), \tag{7}$$

$$g(x_i, t_i) = \sum_{k=1}^{M} \beta_k(t_n)v_k(x_i). \tag{8}$$

We defined $f(x,t)$ as η'_T as thermal steric height anomaly and $g(x,t)$ as SST anomaly (SST') in this study. By determining a regression coefficient between $f(x,t)$ and $g(x,t)$, we estimated a new thermal steric height anomaly from SST'. The detail of the application of the methodology is discussed in *Thermal Steric Height* information in this section.

Sea Level Anomaly (SLA): This series of sea level anomaly (SLA) obtained from a complete reprocessing of TOPEX/POSEIDON (T/P),

Jason and ERS-1/2 data. For the detail information of the SLA, one can refer to the following web site, ftp://ftp.cls.fr/pub/oceano/enact/msla/ readme_MSLA_ENACT.htm. T/P is replaced by Jason in August 2002 after its orbit change. ERS-2 is available from June 1996 to June 2003. It is then replaced by ENVISAT. The SLA was computed using conventional repeat-track analysis. The SLA are relative to a 7-year mean (January 1993 to January 1999). By removing a 7-year mean sea level height, we are able to remove geoid from altimeter measurements. For this study, monthly mean SLA from January 1998 to July 2007 were used. Accordingly, the bathymetric effects on the altimeter measurements are a minimum. A specific processing is performed to get an ERS-1/2 mean consistent with T/P mean. The SLA is provided on a MERCATOR 1/3° grid. Resolutions in kilometers in latitude and longitude are thus identical and vary with the cosine of latitude (e.g. from 37 km at the equator to 18.5 km at 60°N/S).

Thermal Steric Height (η'_T)*:* In order to estimate the sea level variation resulting from thermal expansion in the upper layer, the sea surface height anomaly was calculated using monthly mean XBT (η'_T) data acquired from the Joint Environmental Data Analysis (JEDA) Center. This method integrated temperatures through the water column to 400 m depths from January 1993 to December 2002 using Eq. (3).

In order to achieve fine spatial and temporal resolutions of η'_T, we applied Coupled Pattern Analysis (CPA) between the thermal steric height anomalies derived from monthly mean XBT temperature measurements (1° × 1° grids) and daily Tropical Rainfall Measure Mission (TRMM) Microwave Imager — Sea Surface Temperature (TMI-SST) measurements (1/4° × 1/4° grids) and obtained a merged data set, which was used to calculate the thermal steric height anomalies with a 1/4° resolution, as shown by Yan *et al.* [43]. Then we removed the thermal steric height anomalies from the reprocessed altimetry data and finally obtained weekly salt steric height anomalies with a 1/4° × 1/4° grid resolution.

Sea Surface Temperature (***SST***): In order to obtain better spatial resolution, we computed a new thermal steric height from AVHRR-SST (18 km) after applying an empirical regression curve using CPA as demonstrated in next section. To do that, optimum interpolated sea surface temperature (OISST) with a spatial resolution of a one-degree grid, from January 1993 to December 2002, was also used. The OISST was obtained from the National Oceanic and Atmospheric Administration/National

Centers for Environmental Prediction (NOAA/NCEP) [32]. For the AVHRR (18 km resolution), we obtained them from the Physical Oceanography. Distributed Active Archive Center (PO.DAAC) AVHRR Oceans Pathfinder (ftp://podaac.jpl.nasa.gov/pub/sea_surface_temperature/avhrr/pathfinder/).

Ocean Color Measurements: Chl_a and DAC, the data has been obtained from GES Distributed Active Archive Center (DAAC). The web-site is given by http://daac.gsfc.nasa.gov/data/dataset/SEAWIFS/index.html. The level 3 (L3) Standard Mapped Image (SMI) files that were acquired are image representations of the L3 binned data products. The "bins" correspond to grid cells on a global grid, each cell is approximately 81 square kilometers in size (http://daac.gsfc.nasa.gov/CAMPAIGN_DOCS/BRS_SRVR/seawifsbrs_info.html).

Mooring Observations: In order to validate the estimation of the salinity variation using IMSDA, we used SSS acquired from the Pilot Research Moored Array in the Tropical Atlantic (PIRATA). The web-site is given by http://www.pmel.noaa.gov/pirata/.

Error Analysis:

Tidal aliasing in altimetry: We investigated tidal aliasing in the altimetry measurements. Volkov *et al.* [38] examined tidal aliasing over the northwest European shelf using the different tidal models to correct tidal signals in the altimetry and illustrated that there were high variances between different tidal models over continental shelf. Since the accuracy of altimetry over shallow regions is critical, we referred to Pascual *et al.*'s study ([29], their Fig. 1) for the tidal aliasing in the Yangtze River and the Amazon River outflows. It shows an order of 5 cm root mean square (RMS) error between old (IB, GOT99.2) and new (MOG2D, GOT00.2) corrections. Since there are no direct comparisons between altimetry and tide measurements in the continental shelf of the ECS, it is considered that there are the 20–30% uncertainties in the altimetry. For the Amazon River outflow, we validated salt steric height estimations using the IMSDA far enough offshore in the tropical Atlantic to avoid tidal aliasing in the shallow coastal regions.

Wind induced sea level changes: In order to estimate wind driven sea level height variation over shallow coastal regions to consider the effect of bottom topography, we refer to the quasi-geostrophic and barotropic theory. The relationship between sea level and wind stress can be written as [7].

$$\frac{\partial}{\partial}\nabla^2\eta + \beta\frac{\partial\eta}{\partial x} - \frac{f}{D}\left[\frac{\partial\eta}{\partial x}\frac{\partial D}{\partial y} - \frac{\partial\eta}{\partial y}\frac{\partial D}{\partial x}\right] = \frac{f}{\rho g}[\nabla x(\tau/D)]_z, \qquad (9)$$

where η is the sea surface height; D is the ocean depth; τ is the wind stress; ρ is the seawater density; g is the local gravitational constant; f is the coriolis parameter; and $\beta = df/dy$. Equation (9) relates the sea surface height to the wind stress applied to the ocean surface. In order to keep the balance of Eq. (9), the sea surface height adjusts itself in response to changes of the wind stress forcing. By a scale analysis of Eq. (9), we have

$$\frac{h}{L^2T} \sim \frac{f}{\rho g}\frac{\Gamma}{DL}, \qquad (10)$$

where h is the scale of the sea surface height; L is the spatial scale; Γ is the characteristic wind stress curl; and T is the temporal scale. From Eq. (10), we further obtain

$$h \sim \frac{fTL}{\rho gD}\Gamma. \qquad (11)$$

Equation (11) shows the scale relationship between sea surface height anomaly and the wind stress forcing. For instance, we estimated it over the continental shelf in the ECS. Letting $f = 7.3 \times 10^{-5}\,\text{s}^{-1}$ (at 30°N), $\rho = 1{,}027\,\text{kg}\,\text{m}^{-3}$, $g = 9.8\,\text{ms}^{-2}$, $D = 100\,\text{m}$, $T = 1$ week, $L = 10\,\text{km}$, and $\Gamma = 5 \times 10^{-2}\,\text{Nm}^{-2}$, h can be calculated as 2.3 cm, which is the same order of the accuracy of altimetry for open ocean.

3. Results

3.1. *River discharge*

3.1.1. *Amazon river discharge*

The Amazon River is the largest river system in the world, and contributes about $6 \times 10^{12}\,\text{m}^3$ of freshwater and $1\,\text{Gtyr}^{-1}$ of sediment discharge [6] to the tropical Atlantic. This freshwater is about 16% of the annual discharge into the world's oceans [2]. Hellweger and Gordon [12] studied the Amazon River water into the Caribbean Sea using historical sea surface salinity. However, the estimation of the freshwater outflow was solely dependent on *in-situ* measurements which spatially undersample the region of interest. Consequently, several questions arise, particularly whether we can estimate salinity variation due to the Amazon River discharge into the tropical

Atlantic and whether we can analyze the trajectory of the freshwater after it leaves the North Brazil Current (NBC) using satellite multi-sensor data? The purpose of this case study is to answer the two questions. Specifically, we will demonstrate whether the estimates of variability of freshwater discharge from the Amazon River into the tropical Atlantic can be estimated directly and indirectly using satellite remote sensing data. We considered Chl_a measured by SeaWiFS as indirect evidence of freshwater discharge from the Amazon River, and by the PIRATA mooring data as direct evidence.

Footprint of Freshwater Discharge on Ocean Color Measurements: In order to examine annual variability of salt steric height (η_S), we computed the Standard Deviation (STD) of η_S (Fig. 2). The η_S was integrated from sea surface to 400 m depth using the Levitus 94 climatology data. The STD of these time series is analogous to the long term-times series of the altimetry. Jo *et al.* [14] showed that STD of η_S is more dominant than that of the thermal steric height (η_T). Near the Amazon River mouth, one can see relatively high STD (O (3 cm)) compared to other regions in

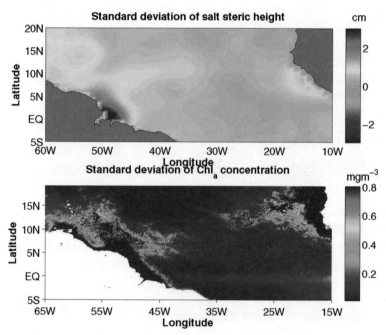

Fig. 2.   Standard deviation of the salt steric height using Levitus 94 in the upper panel, and the standard deviation of the Chl_a in the lower panel using SesWiFs.

the tropical Atlantic (Fig. 2) due to freshwater outflow. Comparing the high STD of η_S in the western tropical Atlantic to the eastern tropical Atlantic, one can see the high STD of η_T off the west coast of Africa: two strong permanent coastal upwelling regions known as the Guinea Dome (12°N, 22°W) and the Angola Dome (10°S, 9°E) as illustrated by Jo *et al.* [14]. How can we identify specific phenomena in the ocean? We discussed how satellite observations can be used to monitor freshwater discharge.

We studied the dispersal of Amazon water with SeaWiFS satellite images to illustrate the extent of front of the Amazon River outflow in the tropical Atlantic. Visible radiance backscattered out of the upper optical depth of the ocean is estimated by SeaWiFS. Nearshore, pigment concentrations are less reliable and are considered only qualitatively. Muller-Karger *et al.* [26] showed that there are high pigment concentrations $(5\,\mathrm{mgm}^{-3})$, which reflect turbidity due to suspended sediment inshore of the 10 m isobath. Figure 2(b) shows the STD of annual ChL_a, which were computed from SeaWiFS data from 2002 to July 2007. Comparing Figs. 2(a) and 2(b) raises the question of whether we can estimate salinity variation using ChL_a or other ocean color measurements. The answer depends on the location of interest, and is largely affected by its distance from land based discharge. Near river mouths, one can successfully estimate salinity using ocean color measurements, but as seen in Figs. 2(a) and 2(b), there are less similarities in salinity and ChL_a off the coast of the Amazon River mouth. Although, there is high biological activity, that does not mean freshwater is present in that region. In fact, high ChL_a off the coast of Africa is due to the upwelling in the Guinea Dome and the Angola Dome; and a possible third signal from the Congo River (5°S, 10°E) discharge.

Validation: The most important tracer of the freshwater discharge in the ocean is the salinity. However, the coarse spatial sampling of salinity measurement throughout the global ocean can present difficulties in analysis. Using Integrated Multi-Sensor Data Analysis (IMSDA), we estimated the salt steric height. In order to validate the Amazon River discharge using $\Delta\eta'_S$, we made a comparison with the mean monthly SSS derived from averaging the daily mean SSS from PIRATA mooring data in Fig. 3. The location 8°N, 38°W, where the North Equatorial Counter Current (NECC) passes through the study region, will be highlighted as an example [16]. This mooring site is the best choice among the PIRATA sites to evaluate low-salinity due to entrainment of the Amazon plume water ($\sim$70%) resulting from NBC retroflection, and due to the NECC's carrying

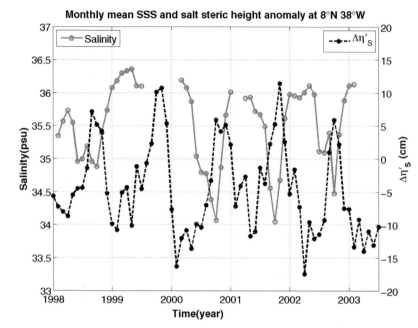

Fig. 3.   Monthly mean SSS obtained from PIRATA mooring station (8°N, 38°W) and salt steric height anomaly derived from Eq. (6). This figure was modified from Fig. 3 in Jo *et al.* [14].

the Amazon plumes eastward into the North Atlantic during summer and fall. The remaining ~30% of the Amazon plume flows northwestward toward the Caribbean Sea [21]. Moreover, by making a comparison at this mooring site, coastal upwelling can be avoided since it is located a large distance from the shoreline. Both time series show that the significant salinity variation occur between October and November in every year. We also found that the precipitation obtained from the PIRATA data does not explain the minimum SSS at this site.

3.1.2. *Yangtze river discharge*

The East China Sea (ECS) is one of the world's largest continental shelf seas and has maintained high levels of primary production and biodiversity. The Yangtze River is the largest river flowing into the ECS, which contributes about 9×10^{11} cubic meters of freshwater and 4.8×10^{11} kilograms of sediment discharge every year. Before the construction of the Three Gorges Dam (TGD), there were many concerns and warnings about its potential

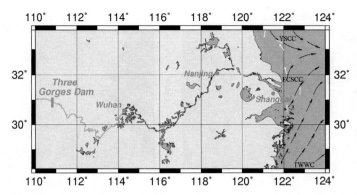

Fig. 4.   The Three Gorges Dam on the Yangtze River and currents in the East China Sea.

negative impacts on the coastal ecosystem, such as a decrease in nutrient-loading and changes in primary production, phytoplankton community structure, and biodiversity of the ECS [42].

The TGD on the Yangtze River and currents in the ECS are shown in Fig. 4. There are three main currents, namely the East China Sea Coastal Current (ECSCC, yellow), the Taiwan Warm Current (TWC, black), and the Yellow Sea Coastal Current (YSCC, black), all of which can affect the distribution of the fresh water discharge from the Yangtze River [9, 10, 28]. In summer, the Yangtze River discharge is driven offshore toward the northeast by the ECSCC. This movement is enhanced by the TWC, which comes from the Taiwan Strait and flows northeastward parallel to the 50 meter isobath, before it turns to east-northeast towards the Korean Strait outside the ECS. The ECSCC meets the YSCC at north of the YRE, where they form a converging area that plays an important role to the distribution of the Yangtze River discharge [9, 10].

Freshwater Outflow Variation Estimated by Salt Steric Height Anomalies: Although there are many published studies on the distribution of chlorophyll a (Chl_a), suspended sediment or yellow substances in the ECS and the YRE using ocean color data observations [11], few studies focus on the impact of fresh water discharge from the Yangtze River. Since, Yan *et al.* [44] demonstrated how well the salt steric height estimated using IMSDA agree with *in situ* salinity measurements in the off coast of YRE, we generate spatial features of the annual mean salt steric height before TGDWS in 2002 and after TGDWS in 2003 (Fig. 5).

It clearly shows that the less the river outflow in 2003, the less the salt steric height. Regarding the salinity contribution to flow fields in the

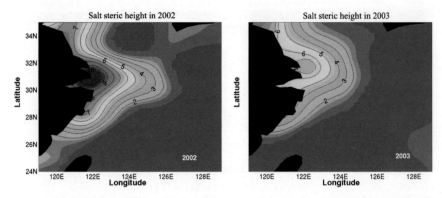

Fig. 5.   Annual mean salt steric height before TGDWS in (a) 2002 and after TGDWS in (b) 2003.

ECS, Yan *et al.* [44] also illustrated how the river outflow (and salt steric height) can change major current systems (Fig. 4). The outflow spreads into the ECS and forms a bulge toward the east due to the convergence of two currents (YSCS and ECSCC). It is worthy to note that there are high uncertainties over the continental shelf in Fig. 5 due to tidal aliasing and wind induced sea surface height. However, the annual mean salt steric height map (Fig. 5) still holds valuable information with respect to spatial features of diluted Yangtze River water in the ECS, although the map does not provide the accurate/absolute salinity variation in the area. Furthermore, the magnitudes of annual mean salt steric height (O (8 cm)) as shown in Fig. 5 near the Yangtze River mouth is slightly higher than errors in altimetry (O (5 cm)) due to tidal aliasing and wind driven sea level changes (O (2 cm)) as discussed in error analysis in Sec. 2. In fact, the amplitude of salt steric height varies from 10 to 18 cm near the river mouth in the ECS [44].

Impact of TGDWS in the YRE: It is interesting to examine how the local current systems changed in the ECS due to TGDWS. Intuitively, we expect less freshwater coming out and thus weaker buoyancy driven flow. For freshwater discharge variation at the YRE, one may refer to the salt steric height anomalies, which was demonstrated by Jiao *et al.* ([13], Fig. S2). In studies by Jo *et al.* [14] and Jiao *et al.* [13], the method of estimating salinity variation from river flow was well demonstrated. In addition to salt steric height anomaly, we considered the effects of both salinity and heat in contributing to changes in the water column. First we computed a time series of salt steric height anomalies near the YRE, which showed the enhanced salinity (or reduced salt steric height anomaly) corresponding to less river discharge and less seasonal variability after TGDWS. Before TGDWS in June 2003, the mean and STD of the salt steric height at 31.75°N 122.0°E were 34.4 cm and 23.6 cm, respectively; while after, they decreased to 26.3 cm and 20.9 cm, respectively [44]. Then we computed thermal steric height anomaly as shown below.

In order to estimate the relative contributions of the two components of steric height anomaly, thermal and saline, we made a steric height anomaly ratio (Rs'), i.e.

$$R'_S = \frac{\eta'_T}{\eta'_S} \approx \frac{\alpha \Delta T'}{\beta \Delta S'}, \tag{12}$$

where α is the thermal expansion coefficient, and the β is the salt contraction coefficient. $\Delta T'$ and $\Delta S'$ are temperature and salinity anomaly differences in the upper layer. The steric height ratio (η'_T/η'_S) is directly related to the density ratio $(\alpha \Delta T/\beta \Delta S)$. Physically, the steric height ratio determines the stability of the vertical water column and thus buoyancy

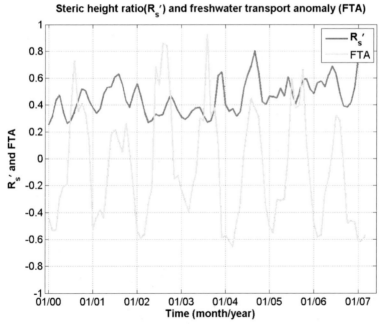

Fig. 6.   Time series of the monthly mean steric height anomaly ratio (R'_s) of smoothed thermal and salt steric height anomaly at 31.75°N 122.0°E. The Freshwater Transport Anomaly (FTA) derived from measurements at Datong Hydrological Station (30.76°N 117.61°E). $3 \times 10^4 \, \text{m}^3$ have to be multiplied to FTA for the original data.

flow field due to density changes. We computed Rs' at 31.75°N 122.0°E in the East China Sea in order to examine the role of freshwater discharge variation, which is critically important in determining the stability of the water column. Before TGDWS in June 2003, the mean and STD of the steric height anomaly ratio (Fig. 6) off coast of YRE were 0.41 and 0.12, respectively, but after that, they increased to 0.47 and 0.19, respectively. The steric height anomaly ratios off coast of YRE showed similar features, which also agreed with the freshwater transport anomaly measured at the Datong Hydrographic Station.

The freshwater transport anomaly was estimated by removing the annual mean from January 2000 to December 2006. The anomalies allow small but significant changes to be more easily observed. The freshwater transport anomaly is positive from June to October in each year, and thus during these months, the mean fresh water transport anomaly before TGDWS (2000 to 2002) is $1.17 \times 10^4 \, \mathrm{m}^4$ with a STD $3.8 \times 10^3 \, \mathrm{m}^3$. After TGDWS from June to October (2003 to 2004), its mean and STD are $1.09 \times 10^4 \, \mathrm{m}^4$ and $2.5 \times 10^3 \, \mathrm{m}^3$, respectively. The freshwater discharge anomaly difference is $0.08 \times 10^4 \, \mathrm{m}^4$.

3.2. Mediterranean eddy and outflow

One of the most interesting and prominent features of the North Atlantic Ocean is the salt tongues originating from an exchange flow between the Mediterranean Sea and the Atlantic through the Strait of Gibraltar. The Mediterranean outflow through the strait is denser than Atlantic water due to its higher salt content. Evaporation in the Mediterranean Sea raises the salinity to around 38.4 psu, compared with 36.4 psu in the eastern North Atlantic. After leaving the strait under the incoming, lighter North Atlantic water, the Mediterranean outflow sinks and turns to the right, due to Coriolis force, following the continental slope off Spain and Portugal. Eventually, this water leaves the coast and spreads into the middle North Atlantic, forming a tongue of salty water. This generates clockwise eddies. These Mediterranean eddies or Meddies [25], as they are often called, are rapidly rotating double convex lenses that contain a warm, highly saline core of Mediterranean water 200–1,000 m thick. Compared to the background water in the Canary Basin, Meddy salinity and temperature anomalies reach 1 psu and 2–4° C higher, respectively [3, 33]. Because of the water differences between a Meddy and the north

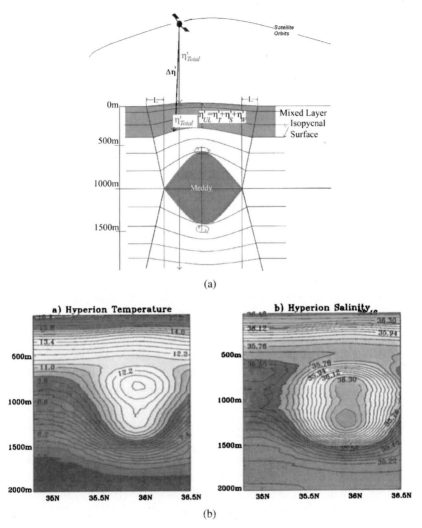

(a)

(b)

Fig. 7. (a) Conceptual diagram of this study based on the isopycnal surface. The aim of this study is to estimate the isopycnal surface at the 400 m depth using mult-sensor data integration analysis. The isopycnal surface at the sea surface is almost flat, however, at 400 m depth there is a significant change that can be detected. (b) This figure is adopted from Tychensky and Carton [37]'s Fig. 5.

Atlantic, isopycnal amplitude reach 6 cm [36, 37]. A Meddy has a typical diameter of approximately 100 km and is located at a depth of 1,000 m. Most Meddy observations have been found in the Canary Basin [1, 15], and in the eastern North Atlantic [3, 31].

How can we observe Meddy and outflow from the Mediterranean Sea in the North Atlantic? A schematic diagram of our method is shown in Fig. 7(a). One can see the relatively broader isopycnal surface (L) as it moves above the Meddy toward the sea surface. Because of this isopycnal compensation, the O&M are not obviously revealed in the η'_{Total} signal at the sea surface. This is why we cannot always detect a Meddy with altimeter data alone. However, the isopycnal surfaces near the Meddy resemble the Meddy's curved structure. Field measurements of the vertical sections of potential density through the three Meddies ([37], Plate 6) showed similar features to those in Fig. 7(a). For comparison between our schematic diagram and *in situ* observation, Fig. 7(b) shows the Hyperion–Meddy observed during *Structures des Echanges Mer–Atmosphere, Proprietes des Heterogeneites Oceaniques: Recherche Experimentale* experiment [37]. The vertical and horizontal structure of temperature and salinity shows that the upper isopycnal surface of the Meddy flattens. However, as we already described, if we can estimate the isopycnal surface below the mixed layer, we may see the existence of a Meddy and/or Meddies. Comparing the isopycnal surface at 1000 m depths, one can see the broader and flatten isopycnal surface at the 400 m depths.

O&M signal and its validation: The result of $\Delta\eta' = \eta'_{\text{Total}} - \eta'_{UL}$ in Fig. 7(a) is apparently some variation of the vertical water column at 400 m isopycnal surface, which is a consequence of the different incoming or outgoing water mass under the upper layer. We can, therefore, investigate the pathways of the O&M in the North Atlantic by using data with better spatial and temporal resolution from space. Yan et al. [43] reported how well O&M signals agree with *in situ* float experiments, *A Mediterranean Undercurrent Seeding Experiment* (AMUSE) [3] and the *Structures des Echanges Mer-Atmosphere, Proprietes des Heterogeneites Oceaniques: Recherche Experimentale* (SEMAPHORE) [33], and XBT measurements during WOCE. In addition to Yan et al.'s [43] annual and monthly comparisons with RAFOS floats, we placed all floats and XBT measurements in seasonal O&M map (Fig. 8). The individual XBT data were obtained from the National Oceanographic Data Center (NODC) in the Global Temperature-Salinity Profile Program database (GTSPP). (More information is available in http://www.nodc.noaa.gov/GTSPP/gtspp-bc.html). While the black dots show the presence of strong O&M, the white dots shows no (or weak) O&M.

The Meddy signals in Fig. 8 are assumed to be the result of multiple Meddies. The condition of the sequential vortice (similarly Meddies)

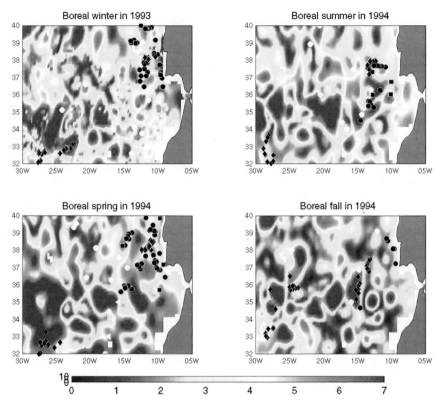

Fig. 8. Seasonal mean signal of the O&M map. Black dots and squares represent Meddies determined by floats and XBTs, respectively. The white dots show weak O&M signals determined from XBTs. The monthly mean O&M map can be also shown in Yan *et al.* [43].

formation was reported by Cenedese and Linden [5] through laboratory experiments. Multi-vortices were generated by a fast flow rate, and a single vortice was generated by a slow flow rate. That is why we can see relatively broader O&M features rather than an individual Meddies. The other reason for broad signals is that this method resolves isopycnal surface at a depth of 400 m (Fig. 7), but not at 1,000 m depth.

Outflow estimation through the Gibraltar Strait: It is important to estimate how much salty Mediterranean water flows over the Strait of Gibraltar for the formation of O&M and thus to assess a salt budget in the north Atlantic Ocean. In Fig. 9, we show the relative transport between a 6°W cross section on the west side of the Strait of Gibraltar and a 5°W

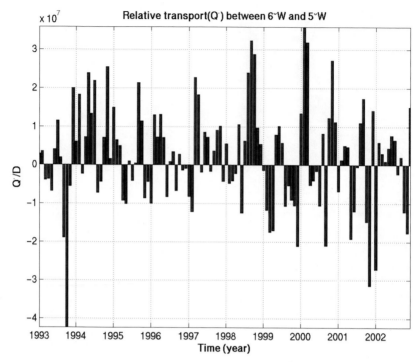

Fig. 9.   Relative transport (Eq. 13) over the Strait of Gibraltar. This figure was modified from Fig. 10 in Yan *et al.* [43].

cross section on the east side of the Strait of Gibraltar,

$$Q' = gD(\eta'|_{x2} - \eta'|_{x1})/f, \tag{13}$$

where g is the gravitational acceleration, D is the depth, f is the Coriolis force, and $\eta'|_{x2}$ and $\eta'|_{x1}$ are the sea level variation at place x_2 and x_1, respectively. Because the exact thickness of the O&M cannot be estimated, we computed Q'/D with respect to time. We can interpret Fig. 9 as follows: if there is more salty Mediterranean water over the Strait of Gibraltar, the salt steric height decreases, and thus decreases total sea surface height. The Mediterranean water can be dense by evaporation in the eastern basin, flows westward below the surface along the west Mediterranean Sea and out through the Strait of Gibraltar into the Atlantic. In Fig. 9, the positive Q' shows that more Atlantic water is transported into the Mediterranean Sea, and the negative Q' shows that more Mediterranean water is transported into the north Atlantic.

4. Conclusions

The IMSDA method is particularly advantageous since satellite multi-sensors overcomes the difficulties of the conventional measurements of freshwater discharge. We believe that this method will help to estimate freshwater and saltier water discharges from rivers, which could be used in determining the freshwater budgets for numerical modeling. However, the IMSDA method has limitations to apply to the coastal regions due to tidal aliasing in the altimetry measurements.

We demonstrated the variability of freshwater discharge from the Amazon River into the tropical Atlantic using SeaWiFS measurements, the IMSDA method, and mooring data. Estimating salinity variation using altimetry is based on the idea that altimetry measures the total change in sea level contributed by the entire water column, whereas XBTs only measures temperature, from which only thermal steric height can be measured. Thus, the effects of temperature and salinity on the total sea level height anomaly can be separated. For the validation, we selected the mooring site at 8°N 38°W to avoid tidal aliasing in the altimetry measurements. The minimum SSS from the PIRATA at 8°N 38°W corresponds well with maximum $\Delta\eta_S$ derived from IMSDA every October.

We investigated the changes in the ECS near the YRE due to TGDWS by comparing salt steric height anomaly before and after TGDWS. Our results show that the salt steric height indicated enhanced salinity and reduced seasonal variability in the area due to less riverine discharge. In addition, the steric height anomaly ratio between heat and salt steric height shows the stability of vertical water column, which can be also used to address the conditions of coastal upwelling in the study area. Accordingly, reduced freshwater outflow is also expected to change surface current patterns in the ECS off the YRE. We expect that this physical parameter, Rs', can be a useful indicator for the prediction/monitoring of ecosystem changes off the YRE.

The IMSDA method was also used to monitor trajectories of O&M. Using this method, we have shown the spatial and temporal variation of the O&M trajectories in the North Atlantic Ocean. The validations were conducted using the AMUSE and the SEMAPHORE experiments and XBT measurements for the vertical temperature distribution. Most of the floats and XBTs' Meddy cases were located in the high signal areas of the O&M map we produced. We estimate relative transport of salty

Mediterranean Water over the Strait of Gibraltar. We found that while more salty Mediterranean water was transported into the north Atlantic in late fall, the Atlantic water was transported into the Mediterranean Sea in early spring over the Strait of Gibraltar.

Regarding application of the IMSDA, it is worthy to note that for the Amazon River and the Yangtze River outflow, we ignored the η'_W (O (2.3 cm)) due to the small contributions to η'_{Alt}. For the Meddy study, we ignored the salinity variation in the upper layer (O (0.12 cm)) and η'_W due to the relatively small (0.5 cm)) compared to η'_{Alt} [43].

Although SSS observations by SMOS (*SMOS* should be launched by ESA) and Aquarius (*Aquarius* is planning to launch in 2010, http://aquarius.gsfc.nasa.gov/) will be available, the IMSDA for estimating salt steric height variation is still important for 3-D geostrophic flow fields and thus climate study.

Acknowledgments

The altimeter products were produced by Ssalto/Duacs and distributed by AVISO, with support from CNES. The SeaWiFS data were provided by the SeaWiFS Project, NASA/Goddard Space Flight Center and ORBIMAGE.

References

1. L. Armi, D. Hebert, N. Oakey, J. Price, P. Richardson, T. Rossby and B. Ruddick, Two years in the life of a Mediterranean salt lens, *J. Phys. Oceanogr.* **19** (1989) 354–370.
2. A. Baumgartner and E. Reichel, *The World Water Balance* (Elsevier, New York, 1975), p. 179.
3. A. S. Bower, L. Armi and I. Ambar, Lagrangian observations of eddy formation during a mediterranean undercurrent seeding experiment, *J. Phys. Oceanogr.* **27** (1997) 2545–2575.
4. C. S. Bretherton, C. Smith and J. M. Wallace, An intercomparison of methods for finding coupled patterns in climate data, *J. Climate* **5** (1992) 541–560.
5. C. Cenedese and P. F. Linden, Cyclone and anticyclone formation in a rotating stratified fluid over a sloping bottom, *J. Fluid Mech.* **381** (1999) 199–223.
6. M. Dagg, R. Benner, S. Lohrenz and D. Lawrence, Transformation of dissolved and particulate materials on continental shelves influenced by large river: Plume processes, *Cont. Shelf Res.* **24** (2004) 833–858.
7. L.-L. Fu and R. A. Davidson, A note on the barotropic response of sea level to time-dependent wind forcing, *J. Geophys. Res.* **100** (1995) 24955–24963.

8. A. E. Gill and P. P. Niiler, The theory of the seasonal variability of the ocean, *Deep-Sea Res.* **20** (1973) 141–177.

9. B. Guan, *Winter Counter-Wind Currents Off the Southeastern China Coast* (China Ocean University Press, 2002), p. 267.

10. B. Guan and G. Fang, Winter counter-wind currents off the southeastern China coast: A review, *J. Oceanogr.* **62** (2006) 1–24.

11. Z. Han, Y.-Q. Jin and C.-X. Yun, Suspended sediment concentrations in the Yangtze River estuary retrieved from the CMODIS data, *Int. J. Rem. Sens.* **27** (2006) 4329–4336.

12. F. L. Hellweger and A. L. Gordon, Tracing Amazon River water into the Caribbean Sea, *J. Mar. Res.* **60** (2002) 537–549.

13. N. Jiao, Y. Zhang, Y. Zeng, W. D. Gardner, *et al.*, Ecological anomalies in the East China Sea: Impacts of the Three Gorges Dam? *Water Res.* **41** (2007) 1287–1293.

14. Y. H. Jo, X.-H. Yan, B. Dzwonkowski and W. T. Liu, A study of the freshwater discharge from the Amazon River into the tropical Atlantic using multi-sensor data, *Geophys. Res. Lett.* **32** (2005) L02605, doi:10.1029/2004GL021840.

15. R. Käse and W. Zenk, Reconstructed Mediterranean salt lens trajectories, *J. Phys. Oceanogr.* **17** (1987) 158–163.

16. E. J. Katz, An interannual study of the Atlantic North equatorial countercurrent, *J. Phys. Oceanogr.* **23** (1993) 116–123.

17. G. E. Lagerloef, Recent progress toward satellite measurements of the global sea surface salinity field, in *Satellites, Oceanography and Society*, ed. D. Halpern, Elsevier Oceanography Series, **63** (2000) 367.

18. G. E. Lagerloef, R. Colomb, D. Le Vine, F. Wentz, S. Yueh, C. Ruf, J. Lilly, J. Gunn, Y. Chao, A. Decharon, G. Feldman and C. Swift, The Aquarius/SAC-D Mission: Special issue on salinity, *Oceanography* **21** (2008) 69–81.

19. G. E. Lagerloef, Final report of the first workshop salinity sea ice working group (SSIWG), preliminary assessment of the scientific and technical merits for salinity remote sensing from satellite (1998), http://www.esr.org/lagerloef/ssiwg/ssiwgrep1.v2.html.

20. G. E. Lagerloef, C. Swift and D. LeVine, Sea surface salinity: The next remote sensing challenge, *Oceanography* **8** (1995) 44–50.

21. S. J. Lentz, Seasonal variations in the horizontal structure of the Amazon Plume interred from historical hydrographic data, *J. Geophys. Res.* **100** (1995) 2391–2400.

22. E. W. Leulitte and J. M. Wahr, Coupled pattern analysis of sea surface temperature and TOPEX/Poseidon sea surface height, *J. Phys. Oceangr.* **29** (1999) 599–611.

23. S. Levitus, R. Burgett and T. P. Boyer, World Ocean Atlas, Vol. 3, Salinity, NOAA Atlas NESDIS 4, U.S. Government Printing Office, Washington, D.C. (1994), 99 p.

24. C. Maes and D. Behringer, Using satellite-derived sea level and temperature profiles for determining the salinity variability: A new approach, *J. Geophys. Res.* **105** (2000) 8537–8547.

25. S. McDowell and H. Rossby, Mediterranean water: An intense mesoscale eddy off the Bahamas, *Science* **202** (1978) 1085–1087.
26. F. Muller-Karger, E. Charles, R. McClain and P. L. Richardsom, The dispersal of the Amazon's water, *Nature* **333** (1998) 56–58.
27. V. S. N. Murty, B. Subrahmanyam, V. Tilvi and J. J. O'Brien, A new technique for the estimation of sea surface salinity in the tropical Indian Ocean from OLR, *J. Geophys. Res.* **109** (2004) C12006, doi:10.1029/2003JC001928.
28. C. E. Naimie, A. B. Cheryl and D. R. Lynch, Seasonal mean circulation in the Yellow Sea — A model-generated climatology, *Cont. Shelf Res.* **21** (2001) 667–695.
29. A. Pascual, C. Boone, G. Larnicol and P.-Y. Le Traon, On the quality of real-time altimeter gridded fields: Comparison with *in situ* data, *J. Atmos. Oceanic Technol.* **26** (2009) 556–569.
30. S. Peng and J. Fyfe, The coupled pattern between sea level pressure and sea surface temperature in the midlatitude North Atlantic, *J. Climate* **9** (1996) 1842–1839.
31. R. Pingree and B. Le Cann, A shallow Meddy (a smeddy) from the secondary Mediterranean salinity maximum, *J. Geophy. Res.* **98** (1993) 20169–20185.
32. R. W. Reynolds and T. M. Smith, Improved global sea surface temperature analyses using optimum interpolation, *J. Climate* **7** (1994) 929–948.
33. P. L. Richardson and A. Tychensky, Meddy trajectories in the Canary Basin measured during the SEMAPHORE experiment, 1993–1995, *J. Geophys. Res.* **103** (1998) 25029–25045.
34. R. W. Schmitt, GOSAMOR, A program for Global Ocean Salinity MonitORing, a proposed contritution to CLIMVAR (1998), http://www.bom.gov.au/bmrc/mrlr/nrs/oopc/godae/gosamor.htm.
35. W. Shi, B. Subrahmanyam and J. M. Morrison, Estimation of heat and salt storage variability in the Indian Ocean from TOPEX/Poseidon altimetry, *J. Geophys. Res.* **108** (2003) 3214, doi:10.1029/2001JC001244.
36. D. Stammer, H. H. Hinrichsen and R. H. Käse, Can Meddies be detected by satellite altimetry? *J. Geophys. Res.* **96** (1991) 7005–7014.
37. A. Tychensky and X. Carton, Hydrological and dynamical characterization of Meddies in the Azores region: A paradigm for baroclinic vortex dynamics, *J. Geophys. Res.* **103** (1998) 25061–25079.
38. D. L. Volkov, G. Larnicol and J. Dorandeu, Improving the quality of satellite altimetry data over continental shelves, *J. Geophys. Res.* **112** (2007) C06020, doi:10.1029/2006JC003765.
39. F. C. Vossepoel, R. W. Reynolds and L. Miller, Use of sea level observations to estimate salinity variability in the tropical pacific, *J. Atmos. Oceanic Technol.* **16** (1999) 1401–1415.
40. J. M. Wallace, C. Smith and C. S. Bretherton, Singular value decomposition of wintertime sea surface temperature and 500-MB height anomalies, *J. Climate* **5** (1992) 561–576.
41. W. Wilson and D. Bradley, Technical Report NOLTR (1966), 66–103 pp.

42. J. Wu, J. Huang, X. Han, Z. Xie and X. Gao, Three Gorges dam — Experiment in habitat fragmentation? *Science* **300** (2003) 1239–1240.

43. X.-H. Yan, Y.-H. Jo, W. T. Liu and M.-X. He, A new study of the Mediterranean outflow, air-sea interactions, and Meddies using multisensor data, *J. Phys. Oceanogr.* **36** (2006) 691–710.

44. X.-H. Yan, Y. Jo, L. Jiang, Z. Wan and W. T. Liu, Impact of the Three Gorge Dam water storage on the Yangtze River outflow into the East China Sea, *Geophys. Res. Lett.* **35** (2008) L05610, doi:10.1029/2007GL032908.

Advances in Geosciences
Vol. 18: Ocean Science (2008)
Eds. Jianping Gan *et al.*
© World Scientific Publishing Company

A TWO-DIMENSIONAL HIGH-RESOLUTION MODEL FOR DEPTH-AVERAGE HYDRODYNAMICS SIMULATIONS IN THE SINGAPORE STRAIT

OLEKSANDR NESTEROV*

Tropical Marine Science Institute,
National University of Singapore
12A, Kent Ridge Rd., 119223, Singapore
* oleksandr.nesterov@gmail.com

ENG SOON CHAN

Department of Civil Engineering,
National University of Singapore
9 Engineering Drive 1,
#07-26, 117576, Singapore

A two-dimensional numerical model for depth-average hydrodynamics simulation in the Singapore Strait is presented. The study has shown that the model is capable to predict hydrodynamics fields fairly well after proper calibration including treatment of the wetting and drying of mangrove areas. It is shown that fine numerical resolution is required in coastal areas not only in the vicinity of those zones, where rapid change of flow magnitude or direction takes place due to shoreline or topography features, but also comparatively far away from them in downstream regions due to possible generation of complicated vortex structures such as von Karman vortex streets.

1. Introduction

A three-dimensional coastal hydrodynamics model needs appropriate vertical resolution taking into consideration both physical and numerical aspects. Coarse vertical resolution often leads to large errors in numerical approximations, especially if a model uses high-order advection-diffusion schemes [9] or turbulence models, such as k/L or k/ε [3]. The latter often requires several tens of layers to ensure convergence and reasonable accuracy of numerical solutions. Having the same computational resources, a two-dimensional model allows finer horizontal resolution, and therefore it may be preferred in those coastal areas, where the impact of complicated shoreline, rapid change in bathymetry, large number of

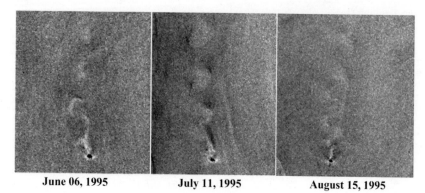

| June 06, 1995 | July 11, 1995 | August 15, 1995 |

Fig. 1.   SAR image of the surface elevation behind the Fairway Rock in the Bering Strait adopted from [10].

narrow rivers and numerous small islands on hydrodynamics fields is more significant than the influence of those factors, which cannot be properly described by two-dimensional equations such as: stratification, large vertical accelerations, considerable variability of flow. Moreover, small islands may cause generation of eddies [4] and von Karman vortex streets, which need to be properly resolved not only in the close proximity to their origins, but also in downstream regions comparatively far away from them. As an example, Fig. 1 adopted from [10] shows perturbations of the surface elevation visible in Synthetic Aperture Radar (SAR) images: vortex street appears behind the island when the mean flow exceeds ∼0.6 m/s. The island has size ∼1 km.

Three-dimensional numerical models applied to simulate currents and water levels in the Singapore Strait one decade ago, such as [16, 17, 18] had rather coarse horizontal resolution of 1 km and used less than 10 vertical layers. Some progress was made in [14] as developments in computing machinery made it possible to refine grid. Parallelization of computations allowed further refinement of horizontal resolution up to 100 m [13]. Though the existence of large eddies in the Singapore Strait was already reported in [14], it can be expected that the presence of many small islands and complicated shoreline may cause generation of smaller-scale eddies and vortex streets.

As currents in the Singapore Strait are strong and water is well-mixed [6], the above mentioned aspects stimulated development of a two-dimensional parallel model for hydrodynamics simulations in the Singapore Strait having finer horizontal resolution of 50 m. It is shown that this model is capable to produce satisfactory results in comparison with measurements proving its applicability to such domains as Singapore Strait. It was found that increased horizontal resolution compared to the previous works reveals generation of many finer-scale eddies and von Karman vortex streets, which are qualitatively similar to those shown in Fig. 1. In addition, it was found that wetting and drying of mangrove areas has significant impact on the flow in the Johor Strait.

2. Model Description

2.1. Governing equations

The model is governed by the following shallow water Eqs. [2, 5]:

$$\frac{\partial \eta}{\partial t} + \frac{\partial(UD)}{\partial x} + \frac{\partial(VD)}{\partial y} = 0, \tag{1}$$

$$\frac{\partial(UD)}{\partial t} + \frac{\partial(U^2D)}{\partial x} + \frac{\partial(UVD)}{\partial y} = -gD\frac{\partial \eta}{\partial x} + \tau_{bx} + \tau_{sx}$$

$$+ fDV + \frac{\partial}{\partial x}\left(A_H D \frac{\partial U}{\partial x}\right) + \frac{\partial}{\partial y}\left(\frac{A_H D}{2}\left(\frac{\partial U}{\partial y} + \frac{\partial V}{\partial x}\right)\right), \tag{2}$$

$$\frac{\partial(VD)}{\partial t} + \frac{\partial(UVD)}{\partial x} + \frac{\partial(V^2D)}{\partial y} = -gD\frac{\partial \eta}{\partial y} + \tau_{by} + \tau_{sy}$$

$$- fDU + \frac{\partial}{\partial x}\left(\frac{A_H D}{2}\left(\frac{\partial U}{\partial y} + \frac{\partial V}{\partial x}\right)\right) + \frac{\partial}{\partial y}\left(A_H D \frac{\partial V}{\partial y}\right), \tag{3}$$

where η — surface elevation, $D = H + \eta$ — total water depth, H — still water depth, U, V — depth-averaged velocity components, (x, y) — Cartesian co-ordinates, t — time. Specific surface shear stress (τ_{sx}, τ_{sy}) is computed basing on prescribed meteorological conditions. Specific bottom shear stress is:

$$\tau_{bx} = -C_d\sqrt{U^2 + V^2}U, \quad \tau_{by} = -C_d\sqrt{U^2 + V^2}V. \tag{4}$$

Bottom drag coefficient C_d is derived in accordance to the classic theory of turbulent boundary layer (see, for example [3]) assuming that vertical profile of velocity magnitude is described by:

$$V(z) = \kappa^{-1}\sqrt{\tau_b} \ \ln(z/z_0), \quad z \geq z_0$$

where z is vertical Cartesian coordinate, $\kappa \approx 0.4$ — von Karman constant, $z_0 = z_0(x, y)$ is the bottom roughness. Then the depth-average magnitude is:

$$\bar{V} = \frac{\kappa^{-1}\sqrt{\tau_b}}{D - z_0} \int_{z_0}^{D} \ln\frac{z}{z_0} dz = \frac{\sqrt{\tau_b}}{\kappa}\left(\frac{D}{D - z_0}\ln\frac{D}{z_0} - 1\right) \approx \frac{\sqrt{\tau_b}}{\kappa}\left(\ln\frac{D}{z_0} - 1\right).$$

It yields the following drag coefficient with additionally imposed minimum:

$$C_d = \max\{0.0025, \kappa^2/(\ln(D/z_0) - 1)^2\}, \qquad (5)$$

Simplified formulation is used in the well-know Princeton Ocean Model [2]:

$$C_d = \max\{0.0025, \kappa^2 \ln^{-2}(D/z_0)\},$$

however, Eq. (5) significantly differs from the latter in shallow regions. Horizontal viscosity coefficient is treated according to the Smagorinsky formula:

$$A_H = C\Delta x\Delta y\sqrt{\left(\frac{\partial U}{\partial x}\right)^2 + 0.5\left(\frac{\partial V}{\partial x} + \frac{\partial U}{\partial y}\right)^2 + \left(\frac{\partial V}{\partial y}\right)^2}, \qquad (6)$$

where $C \sim 0.1 \div 0.2$ is a constant, and Δx, Δy — size of numerical grid cell. Velocity components U and V are set to be zero at rigid lateral boundaries. The time series of the surface elevation η are prescribed along open boundaries together with derivatives of velocity components, which are set to be zero in normal to open boundary direction.

2.2. Numerical realization

Rectangular staggered grids are used for discretization of Eqs. (1)–(6). Optionally the entire computational domain can be split into sub-domains for parallel computations; sub-domains overlapping technique [13] is used for this purpose and MPI (Message Passing Interface) [7] is employed for data exchange between computational nodes.

Advection, diffusion, and Coriolis terms in Eqs. (2) and (3) are discretized explicitly. However, the surface elevation and barotropic forces are computed employing semi-implicit method developed in [5]. The obtained system of linear equations for the surface elevation with five-diagonal positive symmetric matrix is solved by the parallelized Steepest Descent Method with the use of diagonal Jacobi preconditioner [1] to improve convergence.

It was shown [12, 15] that traditional scheme of central differences of the second accuracy order (as in POM, [2]) as well as Lax–Wendroff scheme [9] applied for approximation of the momentum advection terms, may cause spurious oscillations of velocity field in the regions, where flow rapidly (in terms of grid resolution) changes its direction or speed. These oscillations may lead to instabilities and even model failure. Because rapid changes of magnitude or direction are inherent for coastal zones, the above mentioned numerical schemes are often not appropriate. Momentum advection terms are discretized in the presented model with TVD scheme MUSCL adapted for velocity components in staggered grids according to the method developed in [12]. TVD schemes for momentum advection have monotonic behaviour and low dissipation of kinetic energy compared to the 1st-order advection scheme.

Wetting and drying in the model is treated by prohibiting the flow from any "dry" cell, which is defined as a cell where total depth drops below certain threshold value (5 cm). Flow from a "wet" cell to "dry" cell is allowed. Such an algorithm prevents step-change in water level during flooding. Integration time step in the presented model can vary ensuring that CFL condition [9] is always satisfied and that depth is always positive in the areas where drying is possible.

2.3. *Study domain and boundary conditions*

The study domain shown in Fig. 2 covers area approximately 136.2×74.3 km by numerical grid, which consists of 2725×1487 cells and has horizontal resolution 50 m. It is split into sub-domains for parallel computations.

The gridded bathymetry was obtained by spatial interpolation of soundings combined from several commercially-available printed and electronic navigational charts released by Singapore's Maritime and Port Authority (MPA) during the last decade. The maximal depth in the Singapore Strait is about 200 m; however mostly it does not exceed 80 m. The shown bathymetry corresponds to maximal water level, which is 1.6 m above the Mean Sea Level (MSL). The study domain was extended towards the South China Sea as shown in Fig. 2, where the bathymetry is more or less uniform (~ 20 m depth), to prevent development of fictitious circulations at the eastern open boundary.

Large areas in the State of Johor (Malaysia) are covered by mangrove forest and mangrove swaps. Unfortunately, very little information is known regarding topography and physical characteristics of these areas as they

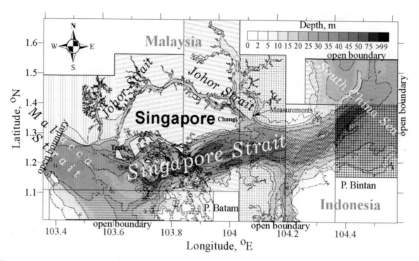

Fig. 2.   The study domain and its splitting into sub-domains for parallel computations.

are covered by forest and hardly accessible for *in-situ* measurements. The extent of mangrove forest was estimated basing on a combination of the MPA admiralty charts and satellite images from Google Earth (www.earth.google.com). Bathymetrical depth in these areas was set according to the closest available values from the admiralty charts; typical range was from 1.5 m below MSL to 1.5 m above MSL. Final adjustments of these areas and bathymetry were made basing on the results of the model calibration.

The model was forced by the time series of surface elevation prescribed at open boundaries of the study domain indicated in Fig. 2. They were obtained using FES'2004 (Finite Element Solution model, http://www.legos.obs-mip.fr) [11], which included 24 harmonics in total, and then linearly interpolated in time and space (along open boundary). Maximal variations of the water level were about ±1.5 m with respect to MSL during spring tide. Discharges in rivers mainly depend on precipitations and were set to be zero for forecast purposes; the use of meteorological and runoff models would be necessary otherwise.

3. Some Results of Application and Discussion

3.1. *Calibration of the model*

Various numerical schemes employed in the model and their realization were verified by respective test cases while the overall model performance

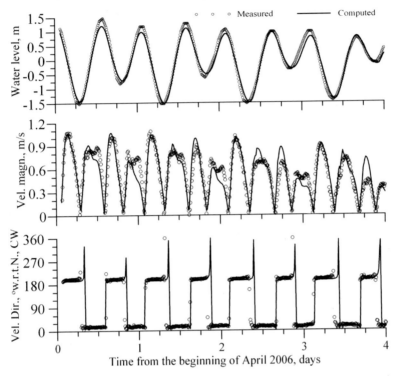

Fig. 3. Comparison of the computed and measured water level, velocity magnitude and direction.

was tested by comparisons with measurements. One of such comparisons of the modeled and observed water levels (with respect to MSL), velocity magnitudes and directions in the Eastern Johor Strait (the location is indicated in Fig. 2) is presented in Fig. 3. Measurements were conducted with Acoustic Doppler Current Profiler (ADCP); the MSL depth at the location was ~17 m; the presented time series is the average over all the ADCP cells. Computations included modeling of wetting and drying. The two-dimensional model was capable to reproduce measured velocity fairly well, proving its applicability. It should be noted that the flow direction was deviating for about 15–20° from the measured direction when a coarser grid with resolution 100 m was used.

The computed velocity fields in the Johor Strait are quite sensitive to the local seabed roughness and presence of areas which can be flooded and dried. The general strategy in the adjustment of local roughness was to achieve the best agreement between the modeled and measured velocity magnitudes. Available observations of the seabed type, such as rocks, gravels, and sand were also utilized. A gridded map of roughness z_0 was created using spatial interpolation. Typical values were set $\sim$3–5 mm in deeper waters of the Singapore Strait and rocky parts of the Johor Strait; $\sim$6–7 mm in areas covered by mangrove forest, and $\sim$1–2 mm in sandy parts of the Johor Strait. In the process of calibration it was found that flow magnitude, inequity of flooding and ebbing durations in the Johor Strait, and presence of multiple velocity peaks during the same tidal phase are likely associated with wetting and drying. This finding was used to make adjustments of mangrove areas, respective bathymetry and roughness in order to achieve the best agreement with the measured flow velocities.

3.2. The influence of wetting and drying on flow in the Johor Strait

Mangrove areas, which are flooded during high tides and dried during low tides, can "absorb" and "release" large amount of water during high and low tides respectively, affecting water flow in the Johor Strait, especially in narrow parts. In general, it is not a trivial problem to parametrize the influence of vegetation and impact of wetting-drying of irregular surface on water flow; modification of all the equations may be necessary as, for example, in [8]. Nevertheless, satisfactory results were achieved with the presented model through adjustments of bathymetry and bottom roughness in the process of the calibration.

To show the influence of wetting and drying on the flow in the Johor Strait two numerical scenarios were computed, one of which included modeling of wetting and drying of mangrove areas (marked as "Computed with W&D") and the other assumed these areas as permanently wet (marked as "Computed without W&D"). In the latter case artificial limit was imposed on the minimal bathymetrical depth, which was set at 1.6 m with respect to MSL. Figure 4 shows the comparison of the measured velocity magnitude during the 2nd day of April 2006 for both of these scenarios. It is obvious that scenario with included modeling of wetting and drying gives better results.

It should be noted that the modeling of wetting and drying of mangrove areas may cause a shift of tidal phase for about one hour: the ebbing becomes shorter, the flooding becomes longer; the moment of time when water level reaches its maximum, is approximately the same for both scenarios and observations. These results are in very good agreement with

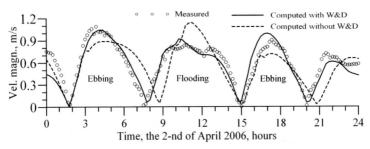

Fig. 4.   Measured velocity magnitude versus "Computed with W&D" and "Computed without W&D". The location of measurements is indicated in Fig. 2.

the measurements and they explain observed inequity of the durations of the flooding and ebbing.

3.3. Predictions of vortices in the Singapore Strait

The increase of horizontal resolution and the use of non-oscillating low-dissipation numerical scheme for approximation of momentum advection allowed modeling of finer-scale eddies behind islands, tips of dams and other shoreline structures. Non-monotonic momentum advection schemes, such as central differences, can be somewhat "stabilized" by the increase of horizontal viscosity coefficient; however such a method does not warranty absence of oscillations and also it causes a side-effect — vanishing of eddies [12].

A sample snapshot of the computed velocity field close to the southern coast of Singapore is shown in Fig. 5. Generation of eddies introduces non-linear effects leading to the appearance of large-magnitude higher-frequency harmonics (in comparison with tidal frequency) in the modeled velocity and even surface elevation. Comparisons with measurements in such regions would be a challenging task due to unstable flow and presence of many fine-scales eddies. It would require simultaneous velocity recording at many locations to understand structure of such flow field. One can also conclude that due to a complex structure of such flow and comparatively long time of eddies existence (compared to the tidal period) it is hardly possible to parametrize their impact on larger-scale transport processes with simple formulations, such as Smagorinsky diffusion coefficient described by Eq. (6).

As simulations have shown, eddies generated behind islands sometimes can be torn away and can be carried downstream by the mean flow, forming von Karman vortex streets. One such vortex street in the eastern Singapore Strait is shown in Fig. 6. Eddies cause noticeable flexures of the surface

144 O. Nesterov and E. S. Chan

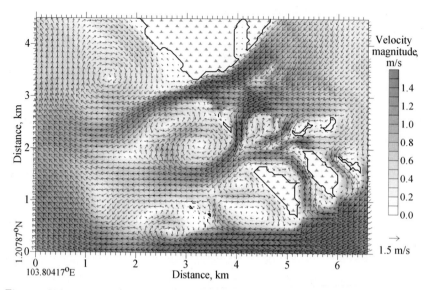

Fig. 5. Velocity field (21 Jun 2008, 20:08): refinement of resolution reveals smaller eddies.

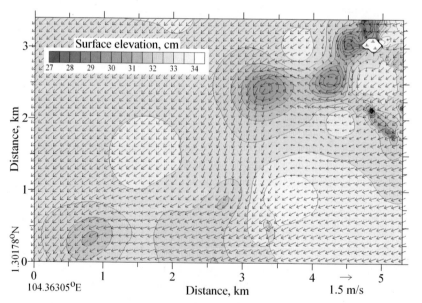

Fig. 6. Example of von Karman vortex street behind island shown as perturbations of the surface elevation (21 June 2008, 22:08).

elevation, qualitatively similar to those visible in Fig. 1. The vortex street appears at this location when flow velocity exceeds $\sim$30–40 cm/s regardless of the direction of the mean flow. The vortex street falls apart into separate eddies which are still noticeable for about 1 hour after the mean flow changes its direction.

Unfortunately, up-to-date SAR images of the Singapore Strait similar to [10] were not available to the authors for direct comparison; however simulations presented in this paper indicate areas which could be a subject of interest for further studies with application of remote sensing technologies. Several smaller vortex streets were identified at the southern coast of Singapore — area, which abounds with small islands; at the northern coast of Indonesian P. Batam $\sim$(104.078°E, 1.205°N). Eddies "rolling" along the coast and causing flow meandering were noticed at the southern coast of Tuas $\sim$(103.623°E, 1.212°N) and at the eastern tip of Changi $\sim$(104.041°E, 1.357°N).

4. Conclusions

Several conclusions have been derived out from this study. Properly calibrated two-dimensional depth-average high-resolution model is capable to predict currents and surface elevation in the Singapore Strait fairly well. Wetting and drying of mangrove areas may significantly affect flow velocity in the Johor Strait and therefore it should be taken into account by the model. High numerical resolution is required not only in the vicinity of those zones, where rapid change of flow magnitude or direction takes place due to shoreline or topography features, but also comparatively far away from them in the downstream regions, where fine-scale flow disturbances can be carried by the mean flow. Promisingly simulated eddies and von Karman vortex streets can be observed in the Singapore Strait with application of remote sensing technologies in the future.

Acknowledgments

The authors are grateful to the National University of Singapore for the support provided during this study. Also the authors would like to express their thanks to two anonymous reviewers, who helped to improve this paper.

References

1. R. Barrett, M. Berry, T. F. Chan, J. Demmel, J. Donato, J. Dongarra, V. Eijkhout, R. Pozo, C. Romine and H. Van der Vorst, *Templates for the Solution of Linear Systems: Building Blocks for Iterative Methods*, 2nd Ed. (Philadelphia: SIAM, 1994), p. 112.
2. A. F. Blumberg and G. L. Mellor, A description of a three-dimensional coastal ocean circulation model, in *Three-Dimensional Coastal Ocean Models*, ed. N. Heaps, *Am. Geoph. Union* **4** (1987) 1–16.
3. H. Burchard and O. Petersen, Models of turbulence in the marine environment — A comparative study of two-equation turbulence models, *J. Marine Systems* **21** (1999) 29–53.
4. R. M. A. Caldeira, P. Marchesiello, N. P. Nezlin, P. M. DiGiacomo and J. C. McWilliams, Island wakes in the Southern California Bight, *J. Geophys. Res.* **110** (2005) C11012.
5. V. Casulli, Semi-implicit finite difference methods for the two-dimensional shallow water equations, *J. Comput. Phys.* **86** (1990) 56–74.
6. E. S. Chan, P. Tkalich, K. Y. H. Gin and J. Obbard, The physical oceanography of Singapore coastal waters and its implications for oil spills, in *The Environment in Asia Pacific Harbours*, ed. E. Wolanski (2006), pp. 393–412.
7. W. Gropp, E. Lusk and A. Skjellum, *Portable Parallel Programming with the Message Passing Interface*, 2nd Ed. (MIT Press, Cambridge, 1999), p. 395.
8. A. Defina, Two-dimensional shallow flow equations for partially dry areas, *Water Resources Research* **36** (2000) 3251–3264.
9. C. Hirsch, *Numerical Computation of Internal and External Flows, Computational Methods for Inviscid and Viscous Flows*, vol. 2 (Wiley & Sons, New York, 1990), p. 714.
10. A. Ivanov Yu and A. I. Ginzburg, Ocean eddies in synthetic radar images, *J. Earth System Science* **111** (2002) 281–295.
11. C. Le Provost, M. L. Genco, F. Lyard, P. Vincent and P. Canceil, Tidal spectroscopy of the world ocean tides from a finite element hydrodynamic model, *J. Geophys. Res.* **99** (1994) 24777–24798.
12. O. Nesterov, Application of TVD schemes for momentum advection approximation in geophysical hydrodynamics models with staggered grids, *J. Appl. Hydromech.* **10** (2008) 58–68.
13. O. Nesterov and P. Tkalich, Hydrodynamics and water quality modeling of Singapore and Johor Strait with SLON model, *Proc. IAHR XXXI Congress*, Seoul, Korea (2005), pp. 4259–4267.
14. W. C. Pang and P. Tkalich, Modeling tidal and monsoon driven currents in the Singapore Strait, *Singapore Maritime & Port J.* (2003), pp. 151–162.
15. G. S. Stelling and S. P. A. Duinmeijer, A staggered conservative scheme for every Froude Number in rapidly varied shallow water flows, *Int. J. Num. Meth. Fluids* **43** (2003) 1329–1354.
16. C. Xiaobo, N. J. Shankar and H. F. Cheong, A three-dimensional multi-level turbulence model for tidal motion, *Ocean Eng.* **26** (1999) 1023–1038.

17. Q. Y. Zhang, E. S. Chan and K. Y. H. Gin, A three-dimensional hydrodynamic model for coastal ocean circulation, *Proc. 2nd OMISAR Workshop on Ocean Models*, Beijing, China (1999).
18. Q. Y. Zhang and K. Y. H. Gin, Three-dimensional numerical simulation for tidal motion in Singapore's coastal waters, *Coastal Eng.* **39** (2000) 71–92.

Advances in Geosciences
Vol. 18: Ocean Science (2008)
Eds. Jianping Gan et al.

LONG TERM MEMORY OF OCEAN DIFFUSION PROCESS IN WEST COAST OF INDIA

YUUKI YAMAMOTO* and MAKOTO KANO
Dept. Electronics and Computer Science,
Tokyo University of Science, Yamaguchi
1-1-1, Daigaku-Dori, Sanyo-Onoda,
Yamaguchi, 756-0884, Japan
*yyama@ed.yama.tus.ac.jp

SHOICHIRO NAKAMOTO
Dept. of Mechanical System Engineering,
Okinawa National College of Technology
905, Henoko, Nago, Okinawa 905-2192, Japan
sn@okinawa-ct.ac.jp

Time series data of current velocities in the Gulf of Kachchh, the west coast of India, were analyzed for quantifying parameters governing coastal water process. Statistical analysis, spectral analysis, and rescaled range analysis were conducted. The velocity diffusion process consists of white noise process and colored noise process. Hurst exponent of time series varies depending on measurement directions implying multi scaling behavior in the coastal water process.

1. Introduction

In predictions of coastal shallow water processes, it is essential to understand time evolutions of velocity distribution function in diffusion processes. Statistical model of velocity diffusion is generally assumed to obey Gaussian process due to its mathematical simplicity. However it is not trivial whether the assumption is applicable to velocity diffusion process in coastal shallow waters.

We have investigated a coastal ocean diffusion process in terms of statistics, periodicity and self affine property. Time series data of coastal ocean current were recorded at hourly intervals for 195 hours by using a moored current meter at near an entrance surface area of the Gulf of Kachchh(GoK) in April, 2002 (See Fig. 1). GoK is known as a

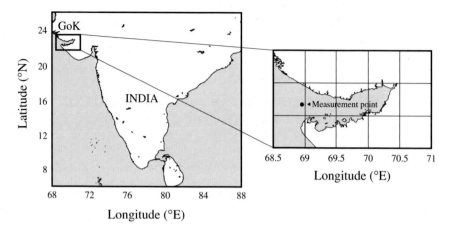

Fig. 1.   Geomorphologic map of Gulf of Kachchh (GoK) and surrounding areas. Time series current data was obtained at entrance area of GoK (filled circle point in right figure).

meso-macro tidal region, large indentation in the northern Arabian Sea coast covering nearly $7000\,km^2$ and has an average depth of about $30\,m$.[1,2] The Tides in the GoK are mixed semi-diurnal type. The shape and orientation of the coast, bathymetry, and funnel shaped geometry are the main reasons for amplification of tides. Currents in the Gulf are mainly tide-driven.

Measurement instrumentation of tidal currents consisted of RCM7 Aanderaa current meter and ancillary apparatus.[a] The kept time to obtain each sampling point of currents was 10 minutes in the measurement. The calculation of tidal currents were implemented using MIKE21 model, including the hydrodynamic(HD) module which integrates equations for mass and momentum conservation. In HD module, major tidal constituents M_2, S_2, K_1, and O_1 were taken into consideration. Trajectories of current velocities in two-dimensional Cartesian coordinates are shown in Fig. 2. Figure 2(a) shows a trajectory of the measured current velocity and (b) shows a trajectory of calculated tidal current velocity. By subtracting calculated tidal current from measured current record, we defined a time series data for the residual tidal current including disturbance signals (Fig. 2(c)).

---

[a] All the measurements of coastal water currents and calculations of tidal currents were conducted by National Institute of Oceanography(NIO), Goa, India.

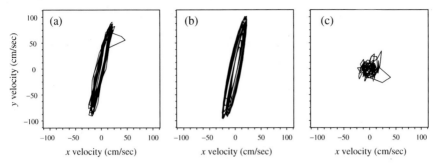

Fig. 2. Trajectories of current velocity. (a) Measured current, (b) calculated tidal current, and (c) residual tidal current.

2. Statistical Characterization of Residual Tidal Current Velocity by Pearson's Chi-squared Test

We have defined radial velocity component $v(t)$ and angular component $\theta(t)$ of residual tidal current velocity in polar coordinates as follows;

$$v(t) = \sqrt{\frac{(v_x - \mu_x)^2}{\sigma_x^2} + \frac{(v_y - \mu_y)^2}{\sigma_y^2}}, \tag{1}$$

$$\theta(t) = \arctan\left(\frac{v_y - \mu_y}{\sqrt{\sigma_y^2}} \Big/ \frac{v_x - \mu_x}{\sqrt{\sigma_x^2}}\right), \tag{2}$$

where μ_x and μ_y are means of residual tidal current velocities in x and y directions, and $(\sigma_x^2)^{0.5}$ and $(\sigma_y^2)^{0.5}$ are standard deviations of residual tidal current velocities in x and y directions, respectively.

Time series data of $v(t)$ and $\theta(t)$ are shown in Fig. 3. A frequency distribution (histogram) of $v(t)$ has been constructed in Fig. 4. A class width of the histogram was chosen according to Scott's rule; $(ClassWidth) = N \log(-1/3)$.[3] The mean and standard deviation of $v(t)$ were estimated to 1.27 and 0.62, respectively. Diamond marks in the figure are for theoretical frequencies sampled from Gaussian (Maxwellian) velocity distribution function in two dimensional polar coordinates given by Eq. (3).[4]

$$f(v) = v \exp\left(-\frac{v^2}{2}\right). \tag{3}$$

To assess a null hypothesis that "the histogram obeys Gaussian distribution in polar coordinates", Pearson's chi-squared test was conducted

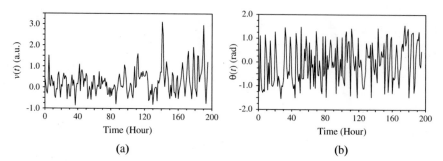

Fig. 3. Time series data of residual tidal current velocity in polar coordinates. (a) Scaled radial component. (b) Angular component.

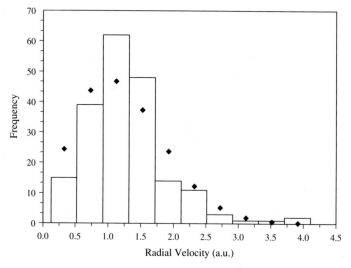

Fig. 4. Histogram of radial velocity in residual tidal current. Diamond marks are for theoretical frequencies (Gaussian in two-dimensional polar coordinates).

using the following equations;

$$\chi_0^2 = \sum_{i=1}^{N} \frac{(O(x_i) - E(x_i))^2}{E(x_i)}, \tag{4}$$

$$g(\chi^2) = \frac{1}{1 - \Gamma\left(\frac{\phi}{2}\right)} \exp\left(-\frac{\chi^2}{2}\right) \left(\frac{\chi^2}{2}\right)^{\frac{\phi}{2}-1}. \tag{5}$$

In Eq. (4), χ_0^2 is a chi-squared value for an observed data, $E(x_i)$ are theoretical frequencies of stochastic variables x_i asserted by the null

Table 1.   Result of Pearson's chi-squared test. Total chi-squared value was estimated to around 42.

Period of radial velocity	Observed frequency	Expected frequency	Chi-sq.
0.329 ~	15	24.410	3.627
0.728 ~	39	43.697	0.505
1.126 ~	62	46.740	4.982
1.525 ~	48	37.319	3.057
1.923 ~	14	23.687	3.961
2.321 ~	11	12.277	0.133
2.720 ~	3	5.270	0.977
3.118 ~	1	1.888	0.418
3.517 ~	1	0.568	0.329
3.915 ~	2	0.144	23.942
	196	196.000	41.932

hypothesis, $O(x_i)$ are observed frequencies, and N is a number of classes in the histogram. Equation 5 indicates the chi-squared distribution, where Γ are the gamma functions with respect to degrees of freedom ϕ. ϕ is set to $N-5$, since parameters of class number N, means (μ_x and μ_y) and variances (σ_x^2 and σ_y^2) in the sample data are estimated in the goodness of fit to Gaussian. Using Eqs. 4 and 5, we estimated χ_0^2 and cumulative chi-squared distribution. The result of the test is shown in Table 1. The estimated χ_0^2 is around 42, so that the cumulative chi-squared distribution ($1-significant$ *level P*) becomes approximately 1, that is, P is approximately 0. Hence the null hypothesis is rejected and it is decided that $v(t)$ does not obey Gaussian distribution in polar coordinates.

3. Power Spectrum and Auto-correlation Analysis

To investigate a memory effect and periodicity in the residual tidal current, auto-correlation and power spectrum analysis were performed. Definitions of auto-correlation $C(\tau)$, Fourier-Laplace Transform $X(f)$ and power spectrum $P(f)$ of stochastic process $x(t)$ are given by the following;

$$C(\tau) = \frac{1}{Z} \sum_t (x(t) - \mu)(x(t + \tau) - \mu), \tag{6}$$

$$X(f) = \sum_{t=0}^{N-1} \exp\left(-i\frac{2\pi}{N}ft\right)(x(t) - \mu), \tag{7}$$

$$P(f) = \frac{1}{N}|X(f)|^2, \tag{8}$$

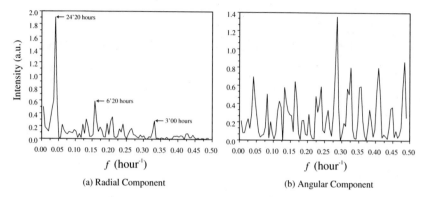

(a) Radial Component   (b) Angular Component

Fig. 5.   Power spectral distributions of residual current velocity. (a) Radial component, (b) angular component. Sampling frequency and Nyquist frequency are 1 hour^{-1} and 0.5 hour^{-1}, respectively.

where Z is the normalization factor, N is a sample number, μ is the mean of stochastic process, f is frequency and $\tau(0 \leq \tau \leq N/2)$ is lag time from criterion time.

Figure 5 shows the power spectra of the radial component and the angular component in the residual tidal current velocity time series data. Before performing the Fourier–Laplace transform of the time series data, window function processing (hamming window) was done. It is observed in Fig. 5(a) that there are sharp peaks at the periods of 24'20, 6'20 and 3'00 hours. Those detected peaks can be due to some memory effects or overestimate (underestimate) of constituent tides in the calculation. On the other hand in Fig. 5(b), there is no outstanding peak. Hereafter, the discussion takes place assuming the calculated tidal currents are correct.

To capture the characteristics of power spectral distributions, the spectra of log-log scale were drawn in Fig. 6. It was observed in Fig. 6(a) that the radial component was of "white noise process" up to about the frequency of $10^{-0.8}$ (hour^{-1}) (6'20 hours). At the frequency range above $10^{-0.8}$ (hour^{-1}), it decreases linearly to $f^{-2.9}$ implying colored noise process. In contrast, the angular component was of "white noise process" at all the frequency range.

Moreover, auto-correlation analysis of each component in the residual tidal current was carried out to investigate velocity relaxation behavior and memory effect (Fig. 7). From Fig. 7(a), it is considered that the auto-correlation of the radial residual tidal current component consists of damping components and sinusoidal components. In literatures, such oscillatory correlation is typically seen in delayed random walks.[5,6] This

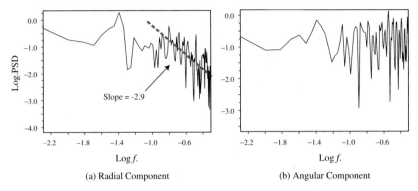

Fig. 6.   Power spectra in log-log scale. (a) Radial component includes white noise process and colored noise process. (b) Angular component is white noise process in the whole frequency.

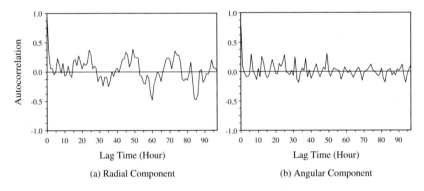

Fig. 7.   Auto-correlations of residual current velocity.

result implies that the radial component of the residual tidal current consists of colored noise and white noise, reflecting long term memory or delay effects. Incidentally, the angular component of the residual tidal current rapidly damped (Fig. 7(b)). This implies the angular component noise has no correlation.

4. Rescaled Range Analysis (R/S analysis)

As stated in Sec. 2, the velocity distribution function of the residual tidal current was not characterized by a conventional Gaussian model. To characterize such non-Gaussian stochastic process, the notion of Brownian motion has to be extended to fractional Brownian motion (fBm).[7] In this section, we perform rescaled range analysis (R/S analysis) to quantify the self-affine scaling behavior of the residual tidal current velocity.[8-10] Using Hurst exponent H, stochastic process $\xi(t)$ is scaled as follows;

$$\xi(t) = \xi(1)t^H. \tag{9}$$

From Eq. 9, moments of $\xi(t)$ are given by

$$\langle \xi(t)^n \rangle = \langle \xi(1)^n \rangle t^{nH} \propto t^{nH}. \tag{10}$$

If the mean of $\xi(t)$ is 0, the variance becomes

$$\langle \xi(t)^2 \rangle \propto t^{2H}. \tag{11}$$

Hence, extent of $\xi(t)$ is linear to t^H (e.g. $H = 0.5$ for standard Brownian motion). An algorithm of R/S analysis is the following;

```
for τ = 1 : N
    mean(ξ(1 : τ));
    ψ(1) = 0;
    for i = 1 : N
        ψ(i + 1) = ψ(i) + ξ(i) − mean(ξ(1 : τ));
    repeat
    R(τ) = max(ψ(1 : τ)) − min(ψ(1 : τ));
    S(τ) = stdev(ξ(1 : τ));
repeat
```

where τ is discrete lagged time, $\psi(\tau)$ denotes cumulative deviation from mean, $R(\tau)$ denotes extent of $\xi(t)(t = 1$ to $\tau)$ and $S(t)$ denotes standard deviation of $\xi(t)$. Plotting $R(t)/S(t)$ versus τ in log-log scale, and fitting them linearly, Hurst exponents ($H = 0$ to 1) can be obtained from the slope of fitting line.

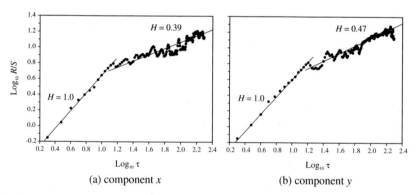

(a) component x (b) component y

Fig. 8.   Results of rescaled range (R/S) analysis. (a) x velocity component, (b) y velocity component. In each of two components, there exist two Hurst exponents per time series.

Results of R/S analysis in x and y directions are shown in Fig. 8. In both of two components, R/S increase linearly to $\tau^{1.0}$ before $\tau = 13$ to 15 hours, and above which both of Hs decrease. R/S in x-direction increases linearly to $\tau^{0.39}$ and R/S in y-direction increases linearly to $\tau^{0.47}$. This result implies that the diffusion process of velocity is not stationary standard Gaussian and has a self-affine fractal property.

5. Conclusions

The following conclusions were derived from what discussed above;

(1) The diffusion process of residual tidal current velocity in the GoK does not always obey ordinary Gaussian distribution.

(2) Spectrum analyses and auto-correlation analyses reveal the studied coastal water process consists of sinusoidal signal and white noise. Sinusoidal behavior of the auto-correlation implies that the residual tidal current has some memory effect.

(3) Hurst exponents (scaling behavior) of x-direction and y-direction are different, that is, the velocity diffusion process has the self-affine fractal property. A possible reason of the difference between x and y directions is due to biased disturbance signals in x-direction from outside sea.

Since the residual tidal current can include ignored deterministic signals, the analysis using an ordinary diffusion equation could not give realistic solutions. Results in this paper can lead to a new analysis method or model of coastal shallow water processes with memory effects. However, it cannot be ignored these conclusions were drawn assuming the calculation of tidal currents by MIKE 21 was correct. It is considered that another method to estimate tidal currents is required to conduct more reliable discussions.

Acknowledgment

We thank Dr. P. Vethamony, National Institute of Oceanography (NIO), Goa, India for providing the processed current meter data.

References

1. P. Vethamony, M. Babua, M. Ramanamurtyb, A. Sarana, A. Josepha, K. Sudheesha, R. S. Padgaonkara and S. Jayakumara, *Mar. Pollut. Bull.* **54** (2007) 697.

2. V. Ramaswamy, B. N. Nath, P. Vethamony and D. Illangovan, *Mar. Pollut. Bull.* **54** (2007) 708.
3. D. W. Scott, *Biometrika* **66** (1979) 605.
4. L. D. Landau and E. M. Lifshitz, *Statistical Physics (Change of Theoretical Physics)*, vol. 5, 3rd ed. (Butterworth-Heinemann, 1984).
5. T. Ohira and J. G. Milton, *Phys. Rev. E.* **52** (1995) 3277.
6. T. Ohira, *Phys. Rev. E.* **55** R1255 (1997).
7. B. B. Mandelbrot and J. W. Van Ness, *SIAM Rev.* **10** (1968) 422.
8. H. E. Hurst, *Trans. Am. Soc. Civ. Eng.* **116** (1951) 770.
9. B. M. Gammel, *Phys. Rev. E.* **58** (1988) 2586.
10. J. L. McCauleya, G. H. Gunaratne and K. E. Bassler, *Physica A* **379** (2007) 1.

Advances in Geosciences
Vol. 18: Ocean Science (2008)
Eds. Jianping Gan et al.
© World Scientific Publishing Company

DYNAMIC TSUNAMI GENERATION PROCESS OBSERVED IN THE 2003 TOKACHI-OKI, JAPAN, EARTHQUAKE

TATSUO OHMACHI

Department of Built Environment, Tokyo Institute of Technology
4259-G3-2 Nagatsuta-cho, Midori-ku, Yokohama 226-8502, Japan
ohmachi@enveng.titech.ac.jp

SHUSAKU INOUE

Department of Built Environment, Tokyo Institute of Technology
4259-G3-2 Nagatsuta-cho, Midori-ku, Yokohama 226-8502, Japan
shusaku@enveng.titech.ac.jp

Using data from the JAMSTEC offshore monitoring system, a dynamic tsunami generation process predicted by the dynamic tsunami simulation technique is verified. The near-field data of the 2003 Tokachi-oki earthquake (M 8.0) is analyzed to show that three kinds of water waves were involved in the process; water pressure waves of short period, tsunami of long period, and water waves induced by Rayleigh waves of intermediate period.

1. Introduction

About a decade ago, the dynamic tsunami simulation technique was developed by Ohmachi et al.[6] to simulate the generation of tsunamis followed by their propagation, where not only the dynamic displacement of the seabed induced by seismic faulting but also the acoustic effects of the seawater are taken into account. In contrast, the conventional method is called the static tsunami simulation technique because of fundamental simplifications employed in the technique. The simplifications lie in the initial sea surface disturbance, which is traditionally assumed to be the same as the static displacement of the seabed caused by seismic faulting, as well as in the propagation of water waves (tsunamis) using the long-wave approximation. Like the aforementioned wording, in this paper, a tsunami generation process associated with dynamic interaction between the seabed and seawater is called a dynamic tsunami generation process.

As the dynamic technique assumes weak coupling between the seabed and the seawater, the simulation consists of a two-step analysis. The first step is the dynamic seabed displacement resulting from a seismic faulting, and the second is the seawater disturbance induced by the dynamic seabed displacement. For the first step, the boundary element method (BEM) is used and, for the second step, the finite difference method (FDM) is used to solve the Navier–Stokes equation. A series of dynamic tsunami simulations have demonstrated that in a far-fault area, the simulated tsunamis have only minor differences between the dynamic and static analyses, but in the near-fault area there are some significant differences such as,

(a) Due to compressibility of the seawater, water pressure change of short period is induced in the dynamic analysis, and lasts longer than input ground motion.

(b) The dynamic seabed displacement induces two types of water waves that travel along the sea surface. One is a long-period tsunami, and the other is a water wave induced by Rayleigh waves arriving prior to the tsunami and having a period shorter than the tsunami.

(c) In the tsunami source area, due to the superposition of two types of water waves, the wave height becomes larger than that resulting from the static simulation.

(d) Several case studies have shown that the dynamic technique has the advantage of predicting near-field tsunamis with higher accuracy.

Although most of these findings look quite reasonable, they have to be verified by observed data (Ohmachi et al.[7]) Hence, in this paper the data obtained by JAMSTEC (Japan Marine Science and Technology Center) for the 2003 Tokachi-oki earthquake is used in the verification of the findings mentioned above.

2. 2003 Tokachi-oki Earthquake and Offshore Monitoring System

The interplate 2003 Tokachi-oki earthquake of JMA magnitude (M_J) 8.0 occurred off the southeastern coast of Tokachi district at 04:50 on September 26, 2003. It was reported that the tsunami arrived at Tokachi Harbor at 04:56 with a wave height of 4.3 m, while the Japanese nation-wide coastal wave observation network (NOWPHAS) located on the 23 m-deep seabed off Ohtsu Harbor recorded the first tsunami arrival at 04:51 (Nagai and Ogawa[4]), which was followed by washing up of fishing boats on to the shore. The locations of these harbors and the epicenter of the earthquake

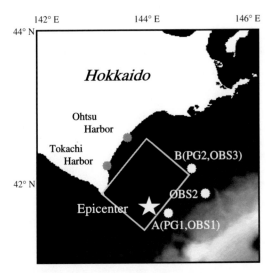

Fig. 1. Locations of the JAMSTEC monitoring system equipment and epicenter of the 2003 Tokachi-oki earthquake.

are shown in Fig. 1, where the location of an offshore monitoring system deployed by JAMSTEC is also shown. The system is equipped with three broadband tri-component seismometers (OBS1-3) and two high-precision pressure gauges (PG1-2). The distances between OBS1 and PG1 and between OBS3 and PG2 are rather short; that is, 4.0 km and 3.4 km, respectively. Thus, as shown in Fig. 1, the locations where these pairs of instruments are installed are respectively referred to as St. A and St. B, and the distance between these two stations is about 72 km. In Fig. 1, the hypocenter of the main shock and a fault model dipping northwest are shown by a star and a rectangle (Koketsu *et al.*[3]). Sampling rates of the pressure gauges and seismometers are 1 Hz and 100 Hz, respectively. Table 1 shows details of the offshore monitoring system and seawater depth of each sensor (JAMSTEC[2]).

3. Tsunami Generation Process Inferred from Observed Data

Time histories of water pressure and ground motion acceleration (vertical component) at St. A and St. B are shown in Fig. 2. At both stations, the water pressure change lasted longer than the ground motion acceleration.

Fourier spectra of the time histories are shown in Fig. 3. Peak periods of 7.0 s at St. A and 6.5 s at St. B are observed for both water pressure and ground motion acceleration, and each period T can be approximated

Table 1. The JAMSTEC offshore monitoring system and water depth of equipment.

Equipment of the system	Depth (m)	Recorded data
Cable-end Station (DSO)	2,540	Flow velocity
		Velocity according to layer
		Hydrophone
A Seismometer (OBS1)	2,329	Ground motion acceleration
		Hydrophone
Tsunami Sensor (PG1)	2,218	Water-Pressure
Seismometer (OBS2)	3,428	Ground motion acceleration
		Hydrophone
B Seismometer (OBS3)	2,124	Ground motion acceleration
		Hydrophone
Tsunami Sensor (PG2)	2,210	Water-Pressure

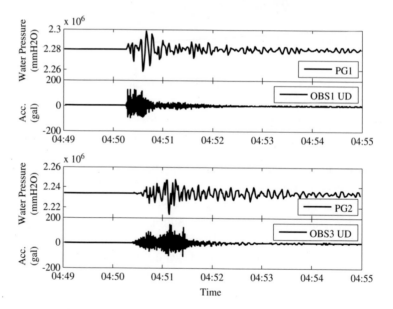

Fig. 2. Time histories of water pressure and vertical ground motion acceleration.

by $T = 4H/c$, where H is the seawater depth $(2.2 \sim 2.3\,\text{km})$ and c is the acoustic wave velocity of water $(1.5\,\text{km/s})$. Hence, these periods are thought to correspond to the acoustic waves traveling between the sea bottom and sea surface (Nosov et al.[5]).

Time histories of the water pressure change shown in Fig. 2 consist of many short period components with amplitudes several times larger than

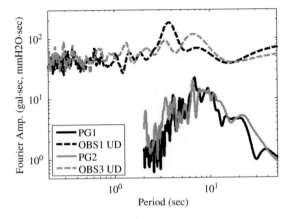

Fig. 3.   Fourier spectra of time histories shown in Fig. 2.

the tsunami-induced water pressure. When the components with period less than 50 s are filtered out, the water pressure time histories are indicated by the gray curves in Fig. 4. It is evident from the gray curves that, after the main shock, the base-line of the water pressure shifted by about 40 cm at PG1 and about 10 cm at PG2 in terms of the water depth. These base-line shifts were supposedly caused by the uplift of the seabed resulting from the seismic faulting (Watanabe *et al.*[8]).

In addition, the gray curves in Fig. 4 indicate two important features. First, a recently generated tsunami was detected with different shapes. At PG2, the tsunami is clearly observed as a solitary wave with a height of about 20 cm and a period of 15 min. At PG1, the wave height and the period of the recently generated tsunami are not as clear as that of PG2, but only a gentle slope can be observed. The difference in the shape of the tsunami is probably due to the difference in the relative location between the observation stations and the tsunami source. As shown in Fig. 1, PG1 is located very close to or just above the seismic fault, while PG2 is located a little apart from the fault and about 72 km away from PG1. Second, another group of water waves preceding the tsunami are seen at the beginning of the short-period water pressure change, and are referred to as preceding waves.

4. Water Waves Induced by Rayleigh Waves

Fourier spectra of the 3-component ground motion acceleration and water pressure at St. A and St. B are shown in Fig. 5. For both St. A and St. B, the spectrum of the NS component has a similarity with that of the UD

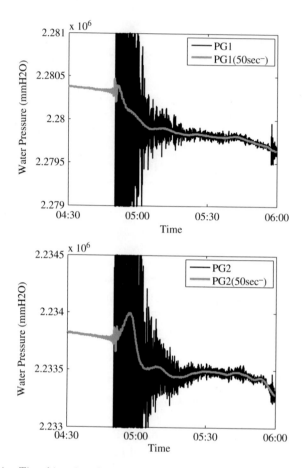

Fig. 4.   Time histories of water pressure after long-pass filtering of 50 s.

component, but not with the ED component. This similarity implies that
seismic ground motions in the NS–UD plane are coupled with each other
and that the motion has a kind of the so-called directivity or directionality.
In addition, according to Fig. 5, for the period ranging from 3 s to 15 s,
both ground motion and water pressure are relatively large in amplitude.

A band-pass filtering between 3 s and 15 s is applied to the original time
histories of water pressure and vertical ground motion displacement at St. A
and St. B and the result is shown in Fig. 6. From Fig. 6, it can be seen that
the water pressure change in this period range shows the largest amplitude
at almost the same time or a little later when the seabed displacement shows
one or two large amplitudes at an initial state of the earthquake motion.

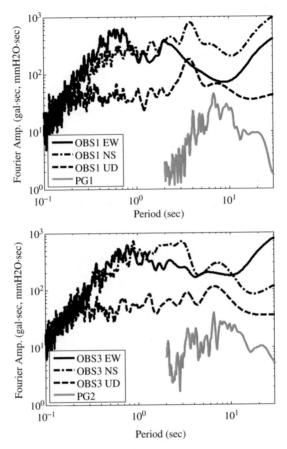

Fig. 5. Fourier spectra of three-component ground motion acceleration and water pressure at St. A (top) and at St. B (bottom).

Figure 7 shows the particle orbit of the ground motion displacement plotted in the NS–UD plane during 4:50:20 and 4:50:40. During the 20 s, the particle orbit had an elliptic and retrograde motion in the NS–UD vertical plane.

From these features of the earthquake ground motion such as the directivity, an elliptic motion on a vertical plane and the predominant period, it seems reasonable to postulate that the water waves observed in the water pressure preceding the recently generated tsunami was induced by Rayleigh waves, as had been predicted from the dynamic tsunami simulation previously. It is noted that as the water waves induced by the Rayleigh waves arrive earlier than tsunamis, its arrival time is likely to lead to a misjudgment of the tsunami arrival time, especially in the near-fault

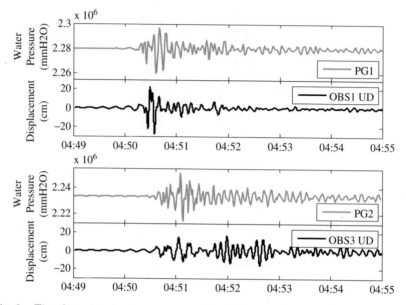

Fig. 6.   Time histories of water pressure and vertical ground motion displacement after band-pass filtering of 3–15 s.

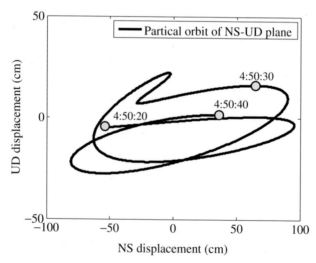

Fig. 7.   Particle orbit of ground motion displacement at OBS1 during 04:50:20 and 04:50:40.

area. In the case of the 2003 Tokachi-oki earthquake tsunami, it seems highly probable that the tsunami arrival time of 04:51 at Ohtsu Harbor was such a misinterpretation.

5. Conclusions

Analysis of the JAMSTEC monitoring data have shown that three kinds of water waves were involved in the dynamic tsunami generation process following the 2003 Tokachi-oki earthquake; that is, water pressure waves of short period, a tsunami of long period, and preceding waves of intermediate period. Their features are summarized in order as,

(a) As for the water pressure waves of short period, the most predominant period was approximately the same as the first mode period T of acoustic oscillation of a water column, expressed by $T = 4H/c$, where H and c are water depth and acoustic wave velocity of the water. Other than this predominant period component, there were several other short-period components. The pressure waves lasted longer than the ground motion and had amplitude larger than that associated with the earthquake-induced seabed displacement, which implies the necessity of proper filtering for detection of the tsunami-induced water pressure.

(b) As for the tsunami of long period, two types of recently generated tsunamis were detected by the water pressure gauges. One was a solitary wave with a height of about 20 cm and a period of 15 min, and the other had a gentle slope and its wave height and period were not as clear as the first type. The first type was detected at a station located a little distance from the seismic fault and the second was detected at a station just above the fault plane.

(c) The water waves preceding the tsunami were detected in an early stage of the water pressure change, and seemed to be induced by a large seabed displacement associated with the Rayleigh waves, because the seismic seabed displacement showed the directivity, an elliptical particle orbit, and a predominant period ranging between 3 s and 15 s. The preceding wave is likely to increase a water wave height when it is superposed on the recently generated tsunami, and lead to misinterpretation of the tsunami arrival time.

The three types of water waves mentioned above were detected by water pressure gauges installed on the seabed and not observed as a water surface disturbance. It is interesting to note that the description by Darwin[1] on the 1835 Chilean earthquake (M 8.0–8.3) refers to the observation of two types of water waves: "In almost every severe earthquake which has been described, the neighboring waters of the sea are said to have been greatly agitated. The disturbance seems generally, as in the case of Conception, to have been of two kinds: first, at the instant of the shock, the water swells high up on the beach, with a gentle motion, and then as quietly retreats; secondly, some little time afterwards, the body of the sea retires from the coast, and then returns in great waves of overwhelming force. The first and less regular movement seems to be an immediate consequence of the earthquake differently affecting a fluid and a solid, so that their respective levels are slightly deranged. But the second case is a far more important phenomenon." Seemingly, the first and second waves in Darwin's description correspond to the preceding wave and tsunami, respectively.

Acknowledgments

The authors are grateful to JAMSTEC for providing the monitoring system data used in this study. The authors would also like to express their thanks to Dr. Anil, C. Wijeyewickrema and Mr. Tetsuji Imai, Professor and Graduate student of Tokyo Institute of Technology, for their kind help in preparation of this paper.

References

1. C. Darwin, *The Voyage of the Beagle* (Penguin Classics, 1989).
2. JAMSTEC, http://www.jamstec.go.jp/scdc/.
3. K. Koketsu, K. Hikima, S. Miyazaki and S. Ide, *Earth Planets Space* **56-3** (2004) 329.
4. T. Nagai and H. Ogawa, Technical note on the Port and Airport Research Institute, vol. 1070 (2004).
5. M. Nosov, S. V. Kolesov, A. V. Ostroukhova, A. B. Alekseev and B. M. Levin, *Doklady Earth Sciences* **404-7** (2005) 1097.
6. T. Ohmachi, H. Tsukiyama and H. Matsumoto, *Bull. Seis. Soc. Am.* **91-6** (2001) 1898.
7. T. Ohmachi, H. Tsukiyama and H. Matsumoto, ITS2001, **5-4** (2001) 595.
8. T. Watanabe, H. Matsumoto, H. Sugioka, H. Mikada and K. Suyehiro, *Eos* **85-2** (2004) 13.

Advances in Geosciences
Vol. 18: Ocean Science (2008)
Eds. Jianping Gan et al.
© World Scientific Publishing Company

CLARIFYING THE STRUCTURE OF WATER MASSES IN EAST CHINA SEA USING LOW-VOLUME SEAWATER MEASUREMENT WITH RARE EARTH ELEMENTS

LI-LI BAI*

*Graduate School of Science and Technology,
University of Toyama, 3190 Gofuku,
Toyama, Toyama, 930-8555, Japan*

JING ZHANG[†]

*Earth and Environmental System,
Graduate School of Science and Engineering for Science,
University of Toyama, 3190 Gofuku,
Toyama, 930-8555 Japan
jzhang@sci.u-toyama.ac.jp*

Rare earth elements in the East China Sea were pre-concentrated by solvent extraction and a back-extraction method with low-volume seawater samples (200 ml). The rare earth element concentrations were measured by connecting inductively coupled plasma mass spectrometry equipment (ICP-MS, ELEMENT II) to a desolvating nebulizer system. Heavy rare earth elements were used to identify the source water masses of the East China Sea, which are composed of Mixed Shelf Water and Kuroshio Current Water; the heavy rare earth elements of all other water samples were at levels between those of these two sources. Kuroshio Current Water is composed of Kuroshio Surface Water, Kuroshio Tropical Water, and Kuroshio Intermediate Water. Heavy rare earth element patterns revealed that the Kuroshio Intermediate Water upwells from a depth of 400 m to depths shallower than 200 m at two sites: in the Okinawa Trough (129.0° E, 31.1° N) and the shelf slope (126.9° E, 28.5° N) of northern of Taiwan Island. The proportions of Kuroshio Intermediate Water were calculated by heavy rare earth elements concentrations in the continental shelf water. From these calculations, Kuroshio Intermediate Water composes 30–40% of the outer-shelf water.

The work was sponsored by
*Japan Science Society, Sasakawa Foundation 19-733M.
†Japan, through the Grants-in-Aid 16681004, 19310007 and 18340143; by the Consignment Study Foundation of Toyama Prefecture.

1. Introduction

The East China Sea (ECS), the largest marginal sea in the western North Pacific and East China regions, is noted for its broad continental shelf. The ECS is mainly fed by four water masses: Taiwan Strait Water (TSW), the Kuroshio Current, the Changjiang River, and the Yellow Sea (YS). Each current has its own direction and is affected by several physical factors. TSW is affected by southwesterly winds in summer (June–August; SW monsoon) and northeasterly winds in winter (December–February; NE monsoon) [1]. Yellow Sea coastal current water starts flowing into the ECS by monsoon in November and stops in March [2]. The Kuroshio Current flows into the ECS from northeast of Taiwan Island. The velocities of the Kuroshio Current vary with depth [3] as well as the upwelling process [4]. In this study, the Kuroshio Current is divided into three water masses: Kuroshio Surface Water (KSW), Kuroshio Tropical Water (KTW), and Kuroshio Intermediate Water (KIW) [5]. KIW is rich in nutrients, and it is distributed widely along the continental shelf. The relative proportion of KIW in the continental shelf water was studied by Chen [6]; however, this proportion varies from place to place. Furthermore, the structures and mixing ratios of the water masses are still unknown due to the complexities of overall water flows and water masses on the continental shelf [7]. For this reason, it is difficult to identify the structures and origins of particular water masses. Different combinations of many hydrographic and chemical data are necessary to clarify these structures and origins.

In this study, rare earth elements (REE) are used as tracers to determine the origins of water masses. The distribution of REE is similar to that of nutrients in that the concentrations increase from surface water to bottom water in the open ocean, making REE useful tracers for determining water mass origin. REE have been measured in the West Pacific Ocean [8] and in the marginal ocean [9]. Shabani *et al.* [10] established the methods used for REE measurement, and they applied a new solvent extraction process for REE extraction. This was later further refined by Zhang *et al.* [11]. The volume of a sample required to measure REE in seawater was 1000 ml [10, 11]. Then, Hongo *et al.* [12] and Wang and Liu [13] reported on the surface water and Changjing estuary water in the ECS, respectively. In these cases, REE were not used for clarifying the structures of water masses in the ECS. However, to apply REE to the assessment of a wide-ranging region, it is necessary to shorten the time needed for sampling collection and to simplify the various analytical processes. To achieve these goals, we developed a new method for REE measurement that uses a low volume of seawater. Moreover, we used REE patterns to clarify the water mass structures in the ECS.

2. Sampling and Analysis

2.1. *Sampling*

A cruise was launched in the ECS on the T/S vessel *Nagasaki-Maru*, from July 9 to 15, 2004. All stations along five sections of the ECS were monitored (Fig. 1). The CTD (SBE-9) observatory was used at all stations to measure temperature and salinity, while water samples for nutrients and other routine analyses were collected vertically with NISKIN bottles at every other station. Furthermore, vertical sampling of REE was carried out at seven stations (expressed by thick solid dots in Fig. 1). In the study area, the average depth at all stations was 200 m, while the deepest was below 1000 m. Most stations were located on the continental shelf. Stations are categorized mainly into three domains: D1 and D2 on the inner-shelf (shallower than 80 m depth); B1–B3, C1–C4, D3–D9, and E1–E6 on the mid-shelf (80–120 m depth); and E7–E8 on the outer-shelf (120–200 m depth). The stations outside of these three domains, with the exception of C9 in the Okinawa Trough, are in the shelf slope.

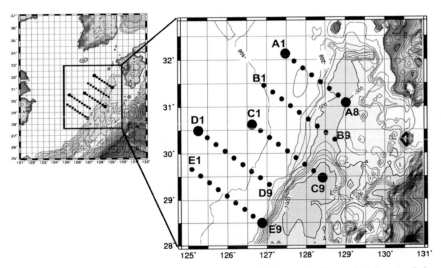

Fig. 1.   Study area and sampling stations in the ECS on the Nagasaki-Maru in July 2004. A CTD observatory was utilized at all stations (thick and small solid dots), thick dots are for REE samples.

2.2. Analysis of hydrochemical parameters

Salinity was measured with the PORTSAL 8410A device (Guideline Instruments Inc.), and the data were used for the calibration of the CTD data. The DO concentration was also calibrated by using the bottle-sample data analyzed by Winkler titration. Nutrients (silicate, nitrate+nitrite, phosphate) were measured using AACS-II (BRAN + LUEBBE), and the limits of detection were silicate: $1 \mu M$, nitrate and nitrite: $0.1 \mu M$, and phosphate: $0.01 \mu M$.

2.3. Low volume measurement of REE

The REE samples were immediately filtered with $0.1 \mu M$ membrane filters within 4 hours after collection. Then, the samples were acidified to a pH below 1.5 using 30% HCl (TAMAPURE-AA-100, Tama Chemicals, Tokyo, Japan) on board. The water samples collected for REE determination were pre-concentrated and analyzed using an inductively coupled plasma mass spectrometer (ICP-MS) (ELEMENT II, Thermo Fisher Scientific K.K.). The ICP-MS ELEMENT II has higher sensitivity and a lower uptake rate than those Plasma Quad ICP-MSs. On the other hand, ELEMENT II experiments involve a higher oxidation rate than that of some elements (e.g. CeO/Ce = 30%), and the produced oxides cause increased interference with other REE. In this study, the ICP-MS was connected to the Aridus Desolvating Nebulizer/Membrane Desolvator (Aridus, CETAC, Technologies, Omaha, Nebraska, USA) to reduce such interactional effects. The features of the Aridus include a low-flow perfluoro alkoxyl alkane (PFA) nebulizer ($100 \mu l/min$ uptake), built-in flow control for Ar sweep gas and additional N_2 gas, and dedicated heat controllers for flexible temperature setting of spray chamber and membrane desolvation unit. The Aridus combines an inert low-flow nebulizer with a highly efficient membrane desolvator to reduce solvent-based interference. Furthermore, the Aridus significantly enhances ICP-MS sensitivity, with improvement of up to 10 times or more. The desolvating membrane of the Aridus significantly reduces sample solvent loading (e.g. H_2O) to the ICP-MS, compared to a conventional pneumatic nebulizer in a chilled spray chamber.

As the results, the CeO/Ce ratio decreased from 30% to 0.2%; other elements' oxidation ((REE)O/(REE)) ratios fell even below 0.1%. In particular, interference between REE pairs, such as $^{140}Ce^{16}O:^{156}Gd$ and $^{151}Eu:^{135}Ba^{16}O$, became lower than 0.01%. The ratios of REE concentration in blanks ($n = 6$) from all pre-concentration procedures, including reagents, to surface waters ($n = 6$; 200 ml) from the Japan Sea were below 4% for [139]La: [146]Nd and 0.6–3% for the other pairings. The precision of surface seawater ($n = 6$) of the Japan Sea were resulted in RSD of 4–5% for La:Pr and 0.3–3% for the others.

As described above, the results indicated that the approach of combining ICP-MS with the Aridus could achieve REE measurement with high-prediction accuracy while using low-volume seawater samples (200 ml). The equipment and analysis conditions are shown in Table 1. In this study, REE concentration measurements of each seawater sample had a volume of 200 ml.

3. Results and Discussion

3.1. Hydrography

All stations are plotted on a temperature-salinity diagram (T–S diagram, Fig. 2), showing density by the solid line in the ranges of 4.00–28.50 in temperature and 31.80–34.90 in salinity. In this figure, the characteristic at station (St.) C9 resembles that of the Kuroshio water, which was reported by Wang *et al.* [12]. Therefore, the characteristic at St. C9 defines the source waters of the Kuroshio Current. Furthermore, the three water masses are defined by temperature: Kuroshio Surface Water along the shelf edge (KSW, T> 25°C), warm and nutrient-poor; Kuroshio Tropical Water (KTW, 25°C > T > 14°C), high salinity; and Kuroshio Intermediate Water (KIW, 14°C >T), cold and nutrient-rich [5, 6]. Furthermore, depth and salinity of these three water masses are defined as follows in this study: KSW

Table 1. Measurement condition of ICP-MS (ELEMENT II) and Aridus.

ICP-MS (Element II)	Cool Gas (N_2) Flow (L/min)	14.80
	Auxiliary Gas (N_2) Flow (L/min)	1.50
	Flush Sample Gas (N_2) Flow (L/min)	0.99
	Fore Vacuum (mbar)	7.11E+04
	High Vacuum (mbar)	2.48E+07
	Argon Middle (bar)	2.50
	Argon Max (bar)	5.04
	Power (W)	1.20E+03
ARIDUS	Sweep Gas Flow (L/min)	5.80–5.90
	Nitrogen Gas Flow (mL/min)	7.00
	Spray Chamber (PFA) Temperature (° C)	110

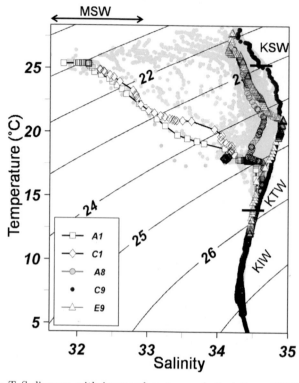

Fig. 2.   The T–S diagram with isopycnal contours of all stations, KSW — Kuroshio
Surface Water; KTW — Kuroshio Tropical Water; KIW — Kuroshio Intermediate Water;
MSW — Shelf Surface Water in the ECS [5, 6].

(above 75 m, $S = 34.80$–34.27), KTW (75 m–370 m, $S = 34.55$–34.90), and
KIW (370 m–1100 m, $S = 34.55$–34.44).

The characteristics of water mass, i.e. salinity and silicate and
phosphate concentrations, differ at a depth of 60–80 m at St. E5. In Fig. 3,
water mass at 60–80 m is higher in salinity (34.50–34.57) and lower in
temperature (18.6–20° C) than the surrounding water. The values of AOU
(30–80 μM), NO_3^- (12.2–8.0 μM), PO_4^{3-} (0.80–0.50 μM), and the ratio of
NO_3^- to AOU (≈ 8) and that of NO_3^- to PO_4^{3-} (≈ 0.08) all agreed with those
of Taiwan Strait Water (TSW) defined by Gong et al. [15, 16]. Moreover,
water masses possessing the same characteristics as TSW were not observed
in other regions. Since St. E5 is in the south of the study area, waters at
St. E5 (TSW) are considered to be a mixture with other water masses.
The earlier study well discussed the Kuroshio water distributed in the ECS.
Consequently, other sources (except Kuroshio water) were defined as Mixed

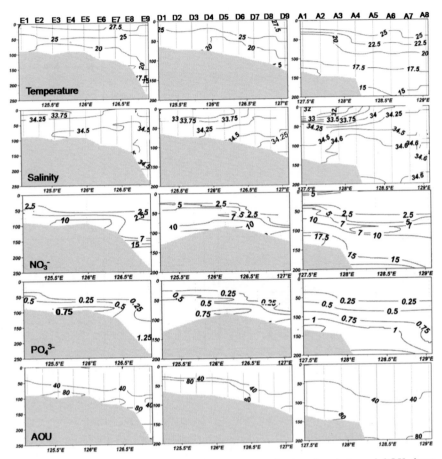

Fig. 3.   Vertical sections of the salinity, temperature, nitrate, phosphate and AOU along transects E (from E1 to E9), D (from D1 to D9) and A (from A1 to A8).

Shelf Water (MSW) in the study area. Therefore, the end members of the ECS are the Kuroshio Current (KSW, KTW and KIW) and MSW, where the salinity of MSW is defined as lower than 33 in this study (Fig. 2).

3.2. Rare earth elements

3.2.1. Distributions of REE concentrations

The distribution of the REE is carried out by a consistent process assumed to be adsorptive scavenging by particles and also possibly by biogenic uptake in coastal areas [17]. However, heavy REE (HREE: Tb, Dy, Ho,

176   L.-L. Bai and J. Zhang

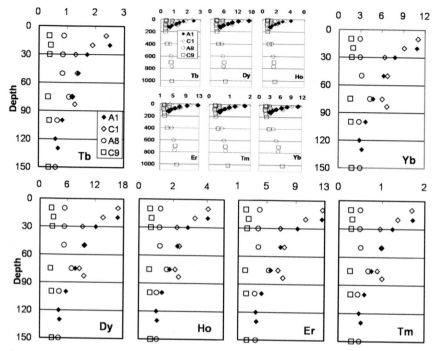

Fig. 4. The vertical profiles of HREE concentrations (Tb, Dy, Ho, Er, Tm, Yb) in the ECS.

Er, Tm, Yb) are minutely affected by these processes compared to the light REE, and thus they are considered a useful tool for investigating the structure of water masses [18]. Nozaki and Alibo [19] reported that HREE scavenging in seawater is small (5%) compared to the case of light REE. Therefore, the HREE are using for studying the varied water sources.

HREE are largely governed by the horizontal processes of ocean circulation with regeneration from particulate matter like dissolved Si [16, 20, 21], their concentrations vary with depth (Fig. 4). HREE have higher concentrations at shallower depths at St. C1 and St. A1, and the concentrations reach a maximum at St. C1 (10 m) and at St. A1 (20 m). Accordingly, the concentrations decreased sharply with depth, from the surface to the bottom. This is because Kuroshio Current water mixes with the MSW, thus diluting the higher concentrations in MSW. Similarly, the minimum HREE concentration is observed at a depth of 75 m at St. C9. However, these concentrations continually increase toward the bottom at St. C9, since REE concentrations increase with greater depth in the open ocean [22].

3.2.2. *Characteristics of water masses by HREE patterns*

To define the water mass structures, REE concentrations were normalized by PAAS (Post Archean Australian Shale) to eliminate the natural Oddo–Harkins effect. In the marine environment, REE patterns have generally been discussed on the basis of shale normalization [16]. In Fig. 5, PAAS normalized HREE patterns of St. A1 (20 m) and St. C1 (10 m), which are defined as MSW ($S < 33$), are displayed in the higher position. The higher REE concentrations are assumed to derive from flows out of Asian rivers suggested by several previous studies [21, 23, 24].

The REE patterns of St. A8 at depths of 75 m and 100 m (Fig. 4) are similar to the patterns of St. A1 at 75 m. Since samples collected at St. A1 contain a mixture of MSW and Kuroshio Current water, we assume that the pattern of St. A8 is strongly influenced by MSW. Therefore, it can be said that MSW flows from the subsurface to St. A8.

On the other hand, the lower-temperature water masses at St. A8 ($<17°$ C) and at St. E9 ($<21°$ C) in Fig. 2 show similar characteristics to those at St. C9. Moreover, the HREE patterns of St. C9 and St. E9 are almost identical at 100 m, which indicates that the two water masses have the same origin. However, the HREE patterns at depths of 150 m and 225 m (bottom) at St. E9 are higher than those at 150 m and 400 m at St. C9 (Fig. 5). This indicates that the upwelling process occurs at St. E9, even though the T–S diagram shows the same values. Similarly, the HREE patterns at a depth of 400 m at St. A8 were plotted in the patterns between

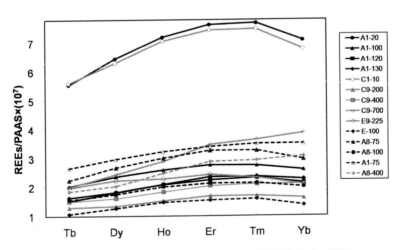

Fig. 5.   The PAAS normalized patterns of HREE in the ECS.

400 and 700 m at St. C9, indicating that the Kuroshio Current is upwelling at St. A8. These results correspond with those of previous studies reporting that the Kuroshio Current has upwelling in the ECS [25].

3.3. Kuroshio intermediate water on outer-shelf

The northernmost St. A1 at a depth from 100 m to the bottom (130 m) has higher salinity (34.55–34.58) and richer nutrients (silicate: 24–25 μM, phosphate: 1.2–1.3 μM). Since the concentrations of nutrients are higher while AOU is the same, the nutrients at a depth of 100 m at St. A1 did not increase due to the decomposition of organic matter. Furthermore, YS, riverine waters, and TSW were unable to provide rich nutrients at St. A1 [26–28]. On the other hand, the HREE patterns at depths of 100 m, 120 m and 130 m at St. A1 were higher than the Kuroshio patterns at 400 m (Fig. 5). This implies that the water mass below 100 m at St. A1 is a mixture of KTW, KIW and MSW. The mixture ratios were calculated according to HREE concentrations (Er, Tm and Yb) and by using the ternary linear equation group. The equations used to make the calculations are as follows.

$$A_1X_1 + B_1X_2 + C_1X_3 = K_1, \tag{1}$$

$$A_2X_1 + B_2X_2 + C_2X_3 = K_2, \tag{2}$$

$$A_3X_1 + B_3X_2 + C_2X_3 = K_3. \tag{3}$$

A: REE concentrations of Er, Tm and Yb in MSW
B: REE concentrations of Er, Tm and Yb in KTW
C: REE concentrations of Er, Tm and Yb in KIW
K: REE concentrations of Er, Tm and Yb at depths of 100 m, 120 m and 130 m in St. A1.

The values of the end members used the averages of MSW (C1: 20 m and A1: 10 m), KTW (C9: 100, 150 and 200 m), and KIW (C9: 400 and 700 m). The results of St. A1 are shown in Table 2, and they reveal that a ratio of 30–40% for KIW is brought onto the outer-shelf. The mixing ratio

Table 2. Mixing ratios of water masses in station A1.

	MSW	KTW	KIW
A1(100 m)	15.3%	55.5%	30.0%
A1(120 m)	4.7%	57.6%	37.7%
A1(130 m)	5.5%	55.1%	40.0%

of KIW below 100 m in the ECS is higher than that described by Chen [6] in all of the continental shelf areas. Consequently, we assume that the KIW is distributed with a lower mixing ratio at the shallower depths (less than 100 m).

4. Conclusions

By connecting the ICP-MS (ELEMENT II) system to the Aridus Desolvating Nebulizer/Membrane Desolvator (CETAC), we successfully made REE measurement with high prediction accuracy using low-volume seawater samples (200 ml). The origins of masses in the ECS, as well as those of MSW, Kuroshio Current water, and mixed waters, were defined by HREE patterns. Flows of KIW in the ECS could be clarified by HREE. KIW at St. E9 and St. A8 experienced upwelling to shallow water along the topography. St. A1 and St. A8 seem to be related due to the high concentration of REE at St. A1, but the relationship has yet to be clarified. Further studies are required to reveal the entire process of supplying nutrients in the region.

Acknowledgments

The authors thank T. Nakamura of University of Toyama for advice on the manuscript, and Prof. T. Matsuno of Kyushu University and X.Y. Guo of Ehime University for many useful comments to improve the discussion. We thank three anonymous referees for their critical comments and valuable suggestions. We thank the captains crewmembers and scientists on the T/S *Nagasaki-Maru* for help with the observations and sampling. We thank N. Yoshinaga of Thermo Fisher Scientific K. K. for help with ICP-MS utilization. This work was supported by the Ministry of Education, Science, Sports and Culture, Japan, through Grants-in-Aid 16681004, 19310007 and 18340143; by the Consignment Study Foundation of Toyama Prefecture; and by the Sasakawa Scientific Research Grant from the Japan Science Society, 19–733M, and supported in part by the Collaborative Research Program of Research Institute for Applied Mechanics, Kyushu University.

References

1. C.-T. A. Chen, M.-C. Yen, W.-R. Huang and W.-A. Gallus, *Monthly Weather Review* **130** (2002) 2271.
2. A. C. Mask and J. J. O'Brien, *J. Geophys. Res.* **103** (1998) 30713.
3. X. Guo, Y. Miyazawa and T. Yamagata, *J. Phys. Oceanogr.* **36** (2006) 2205.

4. C.-T. A. Chen, C.-T. Liu, W.-S. Chuang, Y.-J. Yang, F.-K. Shian, T.-Y. Tang and S.-W. Chuang, *J. Marine Syst.* **42** (2003) 65.
5. S.-L. Wang and C.-T. A. Chen, *Cont. Shelf Res.* **13** (1998) 1573.
6. C.-T. A. Chen, *Oceanologica Acta.* **19** (1996) 523.
7. T. Yanagi, T. Shimizu and H.-J. Lie, *Cont. Shelf Res.* **18** (1998) 1039.
8. E. D. Goldberg, M. Koide, R. A. Schmitt and R. H. Smith, *J. Geophys. Res.* **68** (1963) 4209.
9. J. Zhang and Y. Nozaki, *Geochim.Cosmochim. Acta* **62** (1998) 1307.
10. M. B. Shabani, T. Akagi, H. Shimizu and A. Masuda, *Anal. Chem.* **62** (1990) 2709.
11. J. Zhang, H. Amakawa and Y. Nozaki, *Geophys. Res. Lett.* **21** (1994) 2677.
12. S.-L. Wang, C.-T. A. Chen, C.-H. Hong and C.-S. Chuang, *Cont. Shelf Res.* **20** (2000) 525.
13. Z. L. Wang and C. Q. Liu, *J. Oceanogr.* **64** (2008) 2.
14. Y. Hono, H. Obata, D. S. Alibo and Y. Nozaki, *J. Oceanogr.* **62** (2006) 4.
15. G.-C. Gong, Y.-L.-L. Chen and K.-K. Liu, *Cont. Shelf Res.* **16** (1996) 1561.
16. G.-C. Gong, F.-K. Shiah, K.-K. Liu, Y.-H. Wen and M.-H. Liang, *Cont. Shelf Res.* **20** (2000) 411.
17. H. J. W. D. Baar, M. P. Bacon, P. G. Brewer and K. W. Bruland, *Geochim. Cosmochim. Acta* **49** (1985) 1943.
18. M. Hatta and J. Zhang, *Geophys. Res. Lett.* **33** (2006) L16606, doi:10.1029/2006GL027610.
19. D. S. Alibo and Y. Nozaki, *Geochim. Cosmochim. Acta* **57** (1999) 1957.
20. E. R. Sholkovitz, T. J. Shaw and D. L. Schneider, *Geochim. Cosmochim. Acta* **56** (1992) 3389.
21. D. S. Alibo and Y. Nozaki, *J. Ogeophys.* **105** (2000) 28771.
22. Y. Nozaki and D. S. Alibo, *Geochem. J.* **37** (2003) 47.
23. Y. Nozaki, D. Lerche, D. S. Alibo and M. Tsutsumi, *Geochim. Cosmochim. Acta.* **64** (2000) 23.
24. D. S. Alibo and Y. Nozaki, *J. Geophys. Res.* **105** (2000) 28.
25. T. Yanagi, T. Shimizu and H. J. Lie, *Cont. Shelf Res.* **18** (1998) 9.
26. J. Zhang, S. M. Liu, J. L. Ren, Y. Wu and G. L. Zhang, *Prog. Oceanogr.* **74** (2007) 449.
27. J. Zhang and J. L. Su, Harvard University Press, Cambridge, **14** (2006) 637.
28. D. Y. Gu, L. Z. Chen and S. H. Guo, *Tropic Oceanol.* **11** (1992) 96.

Advances in Geosciences
Vol. 18: Ocean Science (2008)
Eds. Jianping Gan et al.
© World Scientific Publishing Company

ESTIMATION OF GAS HYDRATES AND FREE GAS CONCENTRATIONS USING SEISMIC AMPLITUDES ACROSS THE BOTTOM SIMULATING REFLECTOR

KALACHAND SAIN* and MAHESWAR OJHA

National Geophysical Research Institute,
Uppal Road, Hyderabad 500 606, India
(Council of Scientific and Industrial Research)
* kalachandsain@yahoo.com

Presence of gas hydrates in the marine sediments elevates both P- (V_P) and S-wave (V_S) seismic velocities, whereas even a small amount of underlying free gas decreases the P-wave velocity considerably keeping the S-wave velocity negligibly affected. The amplitude variation with offset (AVO) data from which seismic velocities can be extracted contains useful information for the quantitative assessment of gas hydrate and free gas across the bottom simulating reflector (BSR), an interface between the gas hydrates and free gas bearing sediments. Here we present two techniques based on two completely different approaches to the same multi-channel seismic (MCS) reflection data in the Makran accretionary prism in the Arabian Sea to provide an estimate of gas hydrates and free gas across the BSR at two CDP locations along the seismic profile. In the first approach, we calculate the V_P and V_S and hence the V_P/V_S ratios using the traveltime inversion followed by a constrained amplitude variation with angle (AVA) modeling of the MCS data, and then quantify the amount of gas hydrate as 7–9% and free gas as 3.5–4.0% based on the Biot–Gassmann Theory modified by Lee (BGTL). The second approach, based on AVO intercept (A) and gradient (B) plots, reveals 8–14% gas hydrate underlain by 2.5–5.0% free gas again by using the BGTL model. Both techniques are robust and easy to implement, and the comparable results show an effective strategy for quantifying gas hydrates and free gas across the BSR with more certainty.

1. Introduction

Gas hydrates are formed when low molecular weight guest gas molecules (mainly methane) are trapped within the cage-structure of water molecules at high pressure and low temperature. They are found in the permafrost (130–2,000 m below sediments) and outer continental margins of the world [1–2]. Gas hydrates are attracting the global attention due to their widespread occurrences in nature, potential as a viable source of energy in the 21st century, probable role in global climate change, influence on submarine geo-hazard, and relationship to fluid flow in accretionary wedges. Presence of gas hydrates makes sediments impervious and consequently often traps free gas underneath. Potential reserves of gas in the form of gas hydrates is over $1 - 120 \times 10^{15}\,\mathrm{m}^3$ [3, 4] and is distributed all over the earth at pressure (equivalent to more than 500 m water depth) and temperature less than $12°\,\mathrm{C}$. One volume of methane hydrate contains approximately 164 volumes of methane at normal temperature and pressure. Commercial production of just 15% of gas from gas hydrates would provide the world's energy requirement for the next 200 years at the current level of energy consumption [5].

Quantification of gas hydrates is very important for evaluating the resource potential and assessing the environmental hazards. As the seismic velocity of pure gas hydrates is much higher than that of the host sediments, presence of gas hydrates increases both the V_P and V_S. As the density (920 kg/cc) of pure gas hydrates is almost equal to that of brine, the density of hydrate-bearing and brine-saturated sediments remain almost the same unless there exists a massive hydrate body. On the other hand, free gas below the hydrated sediment reduces the V_P considerably, almost unaffecting the V_S. Though the density of gas is lower than that of brine, it changes very little for low saturation of gas ($< 10\%$) in shallow marine sediments [6]. Thus, we assume constant density for the hydrate- and gas-bearing sediments. The velocity difference between the hydrated sediment above and free gas or brine-saturated sediment below creates the large impedance contrast. This produces large-amplitude and opposite polarity seismic event with respect to the seafloor reflection in MCS data. This reflector is known as the BSR and is most commonly used for the identification of gas hydrates. The BSR typically mimics the shape of seafloor, cross-cuts the dipping sedimentary strata and is often associated with the base of the gas hydrate stability field in thermobaric chemical equilibrium. Changes in various elastic parameters (mainly V_P and V_S) across the BSR can provide vital information in indentifying and estimating gas hydrates and free gas. This change in elastic parameters can be extracted from modeling of seismic amplitudes from the BSR and their differences with respect to the background (without gas hydrates or gas) trend can be used to estimate the saturation of gas hydrates and free gas by employing rock physics modeling.

Using the Zoeppritz equation [7], which governs the AVO/AVA, it is difficult to determine all elastic parameters (V_P, V_S and ρ) across an interface simultaneously [8]. To overcome this difficulty of estimating all elastic parameters, we propose an approach [9], in which V_S of hydrate-bearing sediment is estimated from AVA modeling by constraining the V_P, derived from the Monte Carlo global optimization traveltime inversion. We use another methodology of AVO intercept (A) and gradient (B) crossplot to assess gas hydrates and free gas across the BSR. This approach is widely used in the oil industry for reservoir characterization and lithology identification. The technique also helps understand whether free gas is present below the hydrate-bearing sediments irrespective of saturation of gas hydrates. This approach does not need to estimate V_P and V_S, and avoids error and ambiguities in their estimation. However, A and B are related to V_P and V_S. These two approaches are applied to the MCS data in the Makran accretionary prism. To translate the estimated seismic velocities, and A and B, two rock physics theories are employed for quantitative assessment of gas hydrates and free gas across the BSR.

2. Theories

To interpret seismic data in terms of hydrate content, one needs to establish a relation between hydrate saturation in sediments and their velocity. Lee *et al.* [10] developed a weighted equation (WE) model, in which gas hydrate is considered as part of the pore fluid, although gas hydrates are solid crystal. The velocities predicted from the cementation theory proposed by Dvorkin and Nur [11] are much higher than those normally observed in nature [12]. Helgerud *et al.* [13] introduced an effective medium theory (EMT) that considers the gas hydrates as part of the rock frame and successfully applied their approach to the *P*-wave velocity data from Site 995 of Ocean Drilling Program (ODP) Leg 164 in the Blake Ridge area. Lee [14] proposed a method based on Biot–Gassmann theory (BGTL) to relate the elastic properties of sediments to those of the matrix and the pore fluid. Chand *et al.* [15] compared various models and found modest variations in velocities with hydrate saturation if the same fluid and matrix properties are assumed. Here, we use the EMT and BGTL models to estimate gas hydrate and free gas saturations.

In absence of direct sampling, we assume quartz and clay as the mineralogical constituents for the sandy sediment in the study region [16]. For a seafloor porosity (ϕ_0) of 60% [17] and a compaction constant λ of 1.17 [18], the porosity at the depth of the BSR ($\sim$510 m below the seafloor)

Table 1.   Parameters used in theoretical calculations [14].

Component	K (GPa)	G (GPa)	ρ (kg/cc)
Quartz	36	45	2650
Clay	20.9	6.85	2580
Hydrate	6.41	2.54	910
Methane [20]	0.033	—	112
Water	2.25	—	1000
Critical porosity 36%, No of grain/contact 9, $N = 1$			

is calculated as ~39% using the Athy's law, $\phi(z) = \phi_0 e^{(-z/\lambda)}$ [19]. Other parameters (Table 1) used in the calculation are taken from Lee [14].

3. Field Data

We use a seismic data set in the Makran accretionary prism where the BSR has been identified at about 2.8–2.9 s two way time (TWT) shown in the seismic section (Fig. 1(a)) between CDPs 4275 to 4525. The section is north-south oriented and the CDP number increases landward. A representative normal move out (NMO) corrected CDP gather (Fig. 1(b)) at CDP 4377 with 24 fold and maximum offset of 2559 m shows the reflections from the seafloor and BSR along with the first multiple of seafloor reflection (Fig. 1(c)). The BSR reflection coefficients are calculated from the amplitudes of the seafloor reflection, the BSR and the first multiple of the seafloor reflection using the approach proposed by Warner [21]. The data processing sequence includes a band pass filtering (4–8–50–60 Hz), a spherical divergence correction, minimum phase spiking deconvolution, NMO correction with detailed velocity analysis at about 250 m interval and trace equalization [22]. Root-mean-square amplitudes were picked for a time window of 40–50 ms around the BSR. Spherical divergence correction ($1/(\text{time} \times \text{velocity}^2)$) has been applied with smooth velocity function for less scattering in amplitudes. Array directivity plays a role in AVA analysis [6, 23]. Hydrophone array attenuation is corrected using the function $R = \sin(n\pi(x/\lambda)\sin\theta)/n\sin(\pi(x/\lambda)\sin\theta)$ [24], where n is the number of hydrophones in group, x is the hydrophone interval, λ is the wavelength and θ is the incidence angle. No source directivity correction is applied as the source was of limited spatial extent [16]. As the offset to depth ratio is very small, stretching due to NMO correction is negligible (Fig. 1). Velocity analysis has been carried out many times in an iterative way to get better

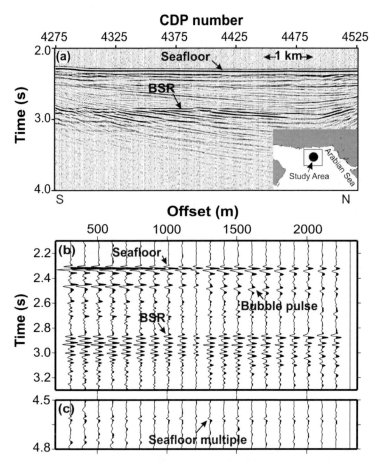

Fig. 1.   (a) Seismic stack section in the Makran accretionary prism (inset shows the location of the study region), (b) a representative NMO corrected seismic gather at CDP 4377 showing reflections from the seafloor and BSR, and (c) the first multiples of the seafloor reflections. CDP increases northward (landward).

stack. The trace to trace or very short wavelength variations are assumed to be noise and need to be eliminated. Trace equalization compensates the amplitude variations due to abnormal shot strength and receiver coupling, etc. But it should be applied very cautiously within a long-gate time window (e.g. seafloor to first multiple of seafloor) at the end of all processing steps [22, 25]. To obtain reflection coefficients with angle of incidence for BSR, offsets are converted into angles as

$$\theta = tan^{-1}(x/z), \tag{1}$$

where θ, x and z are the angle of incidence, half of the offset and depth to the reflector, respectively. The maximum offset of the seismic data is 2.6 km and the depth to the BSR is 2.2 km. For this small offset/depth ratio, the straight path assumption (Eq. (1)), as widely used in seismic data processing by the oil industry and academia, is valid in converting offsets into angles even if a high velocity gradient, due to large concentration of gas hydrate, exists.

4. Methodologies and Application

4.1. *Traveltime inversion and AVA modeling*

Since it is difficult to determine all elastic parameters (V_P, V_S and ρ) across an interface simultaneously from AVA modeling, we perform the constrained AVA modeling [26] and estimate the V_P and V_S of hydrate- and free gas-bearing sediments using the traveltime inversion followed by the AVA modeling of the BSR. The derived V_P/V_S ratio is then matched with the theoretical V_P/V_S ratio for appraising the saturation of gas hydrate and free gas across the BSR.

Since the present approach is based on 1-D modeling, we select the data where the seismic reflectors are almost flat. We perform the traveltime inversion of CDPs 4377 and 4378 in the tau-p domain, and thus transform a supergather of CDPs 4377 and 4378 from the t-x domain to the tau-p domain (Fig. 2) using the method proposed by Korenaga *et al.* [27]. The supergather is formed to avoid the spatial aliasing effect during the transformation of data from t-x to tau-p domain. The advantage of transforming the data into the $tau - p$ domain is that no reflector crosses other reflector(s). Based on the Monte Carlo global optimization technique [28, 29], we derive an optimal P-wave interval velocity-depth model by maximizing the energy along the elliptical trajectories around six major reflectors (marked by arrows in Fig. 2). The reflectors are the seafloor, BSR, one between the seafloor and the BSR and three reflectors below the BSR. We restrict our inversion to six major reflectors, although more are present, because the trade-off between velocities of adjacent thin layers may lead to instability in the inversion if more reflectors are picked. 100 Monte Carlo steps have been carried out for 50 random starting models to reach the global minima followed by few simplex optimization steps to reach the local minima until energy changes by less than one part in 6×10^4. We retain the solution having the maximum energy as the best

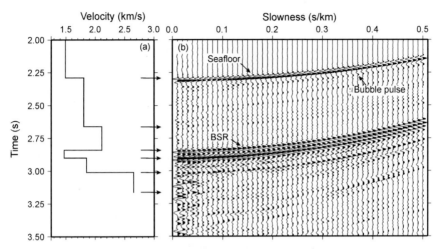

Fig. 2. (a) Velocity-time model derived from traveltime inversion, (b) the tau-p transformed seismic data corresponding to supergather (CDPs 4377 and 4378). Arrows mark the six major reflectors around which traveltime inversion is performed.

solution. The final velocity-depth model is converted into velocity-time model and is displayed in Fig. 2(a) to show the corresponding reflections. The maximum uncertainties in derived P-wave velocities are ±0.08, ±0.17 and ±0.23 km/s for the first three layers, fourth and fifth layers, and the last layer respectively within the 95% confidence limit. The errors in velocities from traveltime inversion propagate downward and increase with depth due to decreasing offset/depth ratio. These errors can be minimized if we have larger offset MCS data. Very accurate velocity can be derived using the sophisticated and computationally intensive full-waveform inversion. However, the errors in velocities associated with the BSR, from which we are estimating saturations of gas hydrates and free-gas, are not much ($\sim 5\%$).

Similarly, we perform traveltime inversion at the supergather of CDPs 4405 and 4406 at a lateral distance of 700 m away from the earlier location. The velocity uncertainties for the first five layers are comparable to those of the first five layers of the earlier velocity model. However, the velocity uncertainty for the sixth layer is ±0.26 km/s. The estimated P-wave velocities for the hydrates- and free gas-bearing sediments across the BSR at these two locations are shown in the Tables 2 and 3. Inversion results show the depth of seafloor lying at 1.70 km and the BSR at 510 m below the seafloor. The increase and drop in velocity across the BSR is due to the presence of gas hydrate and free gas, respectively.

Table 2. Hydrate and free gas saturation corresponding to V_P/V_S ratio across the BSR from BGTL and EMT theories at two locations. S-wave velocities of hydrate bearing sediments, estimated from AVA modeling are shown by bolds. The S-wave velocities of free gas bearing sediments are the background S-wave velocities calculated for 5% clay in BGTL and 25% clay in EMT model, respectively.

CDP supergather	Zone	V_P (m/s)	V_S (m/s) BGTL	V_S (m/s) EMT	V_P/V_S ratio BGTL	V_P/V_S ratio EMT	Saturation (%) BGTL	Saturation (%) EMT
4377–4378	Hydrate	2100	**896**	**881**	2.34	2.38	12	13
	Free gas	1480	825	800	1.79	1.85	4.5	3
4405–4406	Hydrate	2160	**930**	**916**	2.32	2.36	14.5	20
	Free gas	1460	825	800	1.77	1.83	5.5	3.5

Table 3. Corresponding parameters as in Table 2 for the S-wave velocities of free gas bearing calculated for 15% clay in both models incorporating $\pm 3.8\%$ error in P- and S-wave velocities.

CDP supergather	Zone	V_P (m/s)	V_S (m/s) BGTL	V_S (m/s) EMT	V_P/V_S ratio BGTL	V_P/V_S ratio EMT	Saturation (%) BGTL	Saturation (%) EMT
4377–4378	Hydrate	2100 ± 80	$\mathbf{847 \pm 34}$	929 ± 36	2.48	2.26	7	14
	Free gas	1480 ± 56	771 ± 30	860 ± 33	1.92	1.72	3.5	4.5
4405–4406	Hydrate	2160 ± 82	$\mathbf{876 \pm 34}$	969 ± 36	2.46	2.23	9	33
	Free gas	1460 ± 55	771 ± 30	860 ± 33	1.89	1.70	4	5

We calculate the BSR reflection coefficients versus angles for each trace of the CDP gathers between 4373 and 4382 (Fig. 3(a)), and between 4401 and 4410 (Fig. 3(b)), respectively. The reflection coefficients at each angle are averaged out over the said CDP ranges to minimize the errors from field data, and are used for AVA modeling based on the Zoeppritz equation. We perform the constrained AVA modeling [26] and estimate the S-wave velocity for the hydrate-bearing sediments only by keeping fixed the P-wave velocities across the BSR derived from the traveltime inversion, S-wave velocity below the BSR as equal to the background velocity described below and $\rho_2/\rho_1 = 1$. Since the background density is slightly affected by the presence of gas hydrates and free gas across the BSR, the above approximation of uniform density across the BSR is well suited for the AVA modeling from the BSR. It is to be noted here that the BSR is not a geological boundary but a physical contact between gas hydrate- and free gas-bearing sediments. The theoretical responses of the final velocity

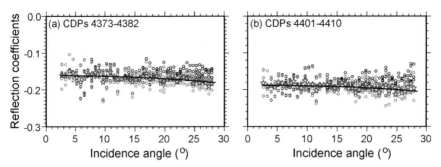

Fig. 3. The reflection coefficients of BSR plotted against the incidence angles for the supergather of CDPs (a) 4373–4382 and (b) 4401–4410 in the Makran accretionary prism. Red and black open circles represent the directivity-corrected and non-corrected reflections coefficients. Blue stars are the average reflection coefficients corresponding to the directivity-corrected data. The black line is the best fit model response (calculated using the Zoeppritz equation) through the average of directivity corrected reflection coefficients.

models (Tables 2 and 3) corresponding to the minimum root-mean-square (rms) residuals between the calculated and observed data are shown in Fig. 3.

In absence of well data, we calibrate our rock physics theories using the derived P-wave velocity and porosity. The P-wave velocity of 1800 m/s, estimated for the layer above the gas-hydrated sediments at 325 m below the seafloor is assumed to be the background velocity, i.e. the sedimentary velocity in absence of gas hydrates and free gas. Using the Athy's law, we calculate the porosity as 45% at this depth. To match the background velocity at 45% porosity, we arrive at 5% clay and 95% quartz based on the BGTL, and 25% clay and 75% quartz based on the EMT model, respectively. Using this fraction of clay and quartz, we calculate the background S-wave velocity at the BSR with 39% porosity based on the BGTL and EMT model, respectively (Table 2). As the presence of free gas negligibly changes the S-wave velocity, the background S-wave velocity is assumed to be the S-wave velocity of the sediment below the BSR. Gassmann's equation [13] is used to calculate the P-wave velocity for different saturations of free gas. The calculated ratios of P to S-wave velocity for models with different saturations of gas hydrates and free gas, respectively, are shown in Fig. 4. We estimate 12% gas hydrate underlain by 4.5% free gas using the BGTL model and 13% gas hydrate underlain by 3% free gas using the EMT model at the CDP supergather of 4377–4378 (Table 2). At other CDP supergather of 4405–4406 (Table 2), we estimate

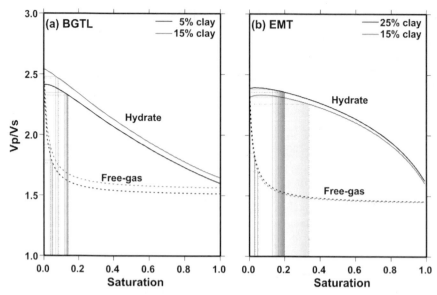

Fig. 4.   Theoretical curves of V_P/V_S ratio versus gas hydrate- (thick solid lines) and free gas- (dashed lines) saturation using (a) BGTL theory for 5% (black lines) and 15% (red lines) clay, and (b) EMT theory for 25% (black lines) and 15% (red lines) clay, respectively. Corresponding to the observed V_P/V_S ratio (Tables 2 and 3), the saturation of gas hydrate and free gas are shown by thin solid and dotted lines for two locations, respectively. Shades between these two lines provide an approximate saturation-limit for gas hydrate and free gas in the study region.

14.5% gas hydrate underlain by 5.5% free gas, using the BGTL model and 20% gas hydrate underlain by 3.5% free gas using the EMT model, respectively.

In reality, the mineralogical constituents at a given location are fixed. For this, we consider 15% clay and 85% quartz as the mineralogical constituents, average of clay and quartz contents predicted by two models, We estimate 7% gas hydrate underlain by 3.5% free gas using the BGTL model and 14% gas hydrate underlain by 4% free gas using the EMT model at the CDP supergather of 4377–4378 (Table 3). At other CDP supergather of 4405–4406, we estimate 9% gas hydrate underlain by 4.0% free gas, using the BGTL model and 33% gas hydrate underlain by 5% free gas using the EMT model, respectively. Even if we have well data, rock physics theories, calibrated with the well data, would also produce different estimates from different rock physics theories. Only some pressure core data could confirm the implication of the use of a particular rock physics theory. To test the influence of the velocity-errors, we perform the

AVA/AVO modeling by incorporating $\pm 3.8\%$ error (according to $\pm 80\,\text{m/s}$ error in $2100\,\text{m/s}$ of P-wave velocity across the BSR at CDP 4377–4378) in both P- and S-wave velocities across the BSR. The results (Table 3) shows almost the same V_P/V_S ratios and hence the same estimations. This is due to the Zoeppritz equation, which gives same impedance contrast for the same V_P/V_S ratio. But if we incorporate the error in P-wave velocity only, the saturation estimations differ a lot. This means that the V_P/V_S ratio technique is more useful in estimating saturations of gas hydrate and free gas than that of using the V_P or V_S alone.

4.2. *AVO crossplot*

In the hydrocarbon industry, AVO attributes from lithologic boundaries have gained considerable popularity for predicting the lithology and reservoir characteristics [30, 31]. Several authors [6, 12, 23, 32, 33] have estimated gas hydrates and free gas saturations across the BSR using conventional AVO modeling coupled with various rock physics models. To overcome the non-uniqueness involved in AVO modeling, we present an approach of estimating the AVO intercepts (A) and gradients (B) from the BSR and comparing them with the theoretical results for several gas hydrate models to quantify the saturation of gas hydrates and free gas across the BSR. Conventionally, the AVO technique involves first the estimation of reflection coefficients, then the estimation of elastic parameters from AVO modeling/inversion, and finally estimation of the amount of gas hydrates and free gas based on a rock physics modeling. In the present approach, we skip the second step and thus avoid some of the uncertainties involved in deriving the elastic parameters [34]. For a plane wave incident at an interface between two semi-infinite isotropic homogeneous elastic half-spaces, and for angles of incidence θ up to $\sim 30°$, the P-wave reflection coefficient $R(\theta)$ can be expressed [35] through the AVO intercept (A) and the AVO gradient (B) as

$$R(\theta) \approx A + B \sin^2 \theta, \tag{2}$$

where

$$A = \frac{1}{2}\left(\frac{\Delta V_P}{V_P} + \frac{\Delta \rho}{\rho}\right), \tag{3}$$

$$B = A A_0 + \frac{\Delta \sigma}{(1-\sigma)^2}, \tag{4}$$

$$A_0 = B_0 - 2(1 + B_0)\frac{1 - 2\sigma}{1 - \sigma}, \quad B_0 = \frac{\Delta V_P / V_P}{\Delta V_P / V_P + \Delta \rho / \rho}$$

$$V_P = (V_{P2} + V_{P1})/2, \quad V_S = (V_{S2} + V_{S1})/2,$$

$$\rho = (\rho_2 + \rho_1)/2, \quad \Delta V_P = (V_{P2} - V_{P1})$$

$$\Delta V_S = (V_{S2} - V_{S1}), \quad \Delta \rho = (\rho_2 - \rho_1),$$

$$\Delta \sigma = (\sigma_2 - \sigma_1), \quad \sigma = (\sigma_2 + \sigma_1)/2$$

and

$$\sigma_i = (5V_{Pi}^2 - V_{Si}^2)/(V_{Pi}^2 - V_{Si}^2).$$

The subscripts 1 and 2 represent the parameters of overlying and underlying layers, respectively. Equations (2)–(4) are valid up to high A for a negative impedance contrast across an interface [31, 26], and to about $A = 0.2$ for a positive impedance contrast.

Considering BSR as an interface, first we compute A and B attributes using Eqs. (2–4) by varying the saturation of gas hydrate from 0 to 80% at an interval of 5% above and 0% free gas (or 100% brine) below the BSR. In a similar way, next we compute A and B attributes by varying the saturation of gas hydrate from 0 to 80% at interval of 5% for fixed saturations of 1, 2, 5 and 10% free gas, respectively. Above $\sim 10\%$ gas saturation, the P-wave velocity remains almost unchanged for uniform distribution of gas. The nonlinear behavior of A-B crossplots for various gas hydrate models are displayed in Fig. 5. The A-B crossplot for hydrates/brine BSR falls in the 2nd quadrant (negative A and positive B) and is treated as the reference trend. In presence of free gas, the A-B crossplots are clearly deviated from the reference. Thus, the A-B crossplot can be used as an indicator of free gas below the hydrated sediments, irrespective of gas hydrates saturation and consequently as a tool for the estimation of both gas hydrates and free gas saturation.

We calculate the intercept A and B (Fig. 6) for two sets of CDPs: 4373–4382 and 4401–4410 using Eq. (2). To get reflection coefficients with angle of incidence, offsets are converted into angles using Eq. (1). A and B values are superimposed on Fig. 5. As the reflection coefficients are scattered, we have estimated probable limits of gas hydrate and free gas saturations considering lower and upper bound fits. The A-B attributes corresponding to the BGTL model indicate 8–14% gas hydrate underlain by 2.5–5.0% free gas, respectively, whereas, the A-B attributes corresponding to the

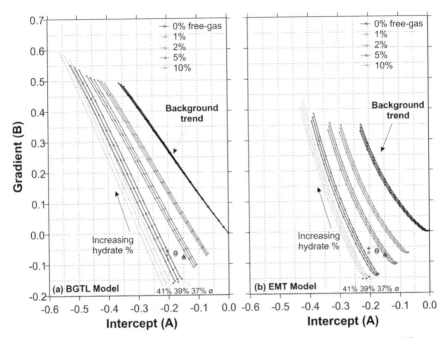

Fig. 5. Theoretical crossplots between intercept (A) and gradient (B) for different saturations of gas hydrates with several fixed saturation of free gas at 39±2% porosity (ϕ) using (a) BGTL and (b) EMT, respectively. The crosses, circles and triangles represent the *A-B* values for lower bound, best fit and upper bound lines corresponding to the $R(\theta)$ vs $\sin^2(\theta)$ curves (Fig. 6), respectively, for various CDP ranges of 4373–4382 (blue) and 4401–4410 (red) respectively.

EMT model indicate 20–27% gas hydrates underlain by 2.0–3.5% free gas, respectively.

We have incorporated ±5% error in porosity estimation, i.e. 39 ± 2% porosity to see the effect on saturation of gas hydrates and free gas. From the present study, we observe that the estimation of gas hydrate saturation is more sensitive to porosity in EMT than in BGTL model. The BGTL model shows almost linear trends with gas hydrate and free gas saturations, whereas the EMT model is nonlinear with gas hydrate and free gas saturation, and even with porosity (Fig. 5).

5. Discussion and Conclusions

We present two AVO methods based on two rock physics theories for the estimation of gas hydrates and free gas across a BSR. The *P*-wave velocity

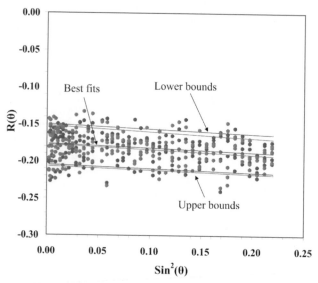

Fig. 6. BSR reflection coefficients, $R(\theta)$ vs $\mathrm{Sin}^2(\theta)$ plots for CDP gathers of (i) 4373–4382 (blue) and (ii) 4401–4410 (red). The lower bound, best fit and upper bound lines corresponding to the $R(\theta)$ vs $\mathrm{Sin}^2(\theta)$ curves are also shown.

from traveltime inversion and the S-wave velocity from AVA modeling is an important strategy to determine both the P- and S-wave velocities from the MCS data and hence estimate the saturation of gas hydrates and free gas across the BSR. We estimate 7 to 9% gas hydrate and 3.5 to 4.0% free gas, respectively, using the BGTL model at two CDP locations with 700 m lateral distance. Whereas, based on the EMT model, we predict higher values of gas hydrate (14 to 33%) and free gas (4.5 to 5.0%), saturation, respectively.

The crossplot of AVO intercept (A) and gradient (B) detects free gas irrespective of hydrates saturations and estimates both gas hydrates and free gas across the BSR. The A-B attributes corresponding to the BGTL model indicate 8 to 14% gas hydrate underlain by 2.5 to 5.0% free gas, respectively, northward along the seismic line. Whereas, the EMT model indicates 20 to 27% gas hydrate underlain by 2.0 to 3.5% free gas, respectively, laterally. This approach skips the estimation of elastic parameters (like V_P and V_S) and thus avoids some uncertainties involved in deriving them. The technique is robust and easy to implement. From this study we conclude that different rock physics theories give rise to different estimates for a given mineralogical constituents and porosity. Therefore, only the pressure core data can validate the use of a particular rock physics theory.

Although there are certain errors in determining P-wave velocities and reflections coefficients, which causes errors in estimation, both methods give an overview of concentrations along the seismic profile and reveal increasing saturations towards north (landward). In a qualitative way, we have included the effects of change in porosity and mineralogy on the saturations of gas hydrates and free-gas. The potentiality of the methods can be established when well data are available

It is observed from Fig. 5(b) that reflection coefficients are slightly negative at low hydrate saturations ($< 10\%$) and 0% free gas. The V_P/V_S ratio increases with hydrate saturation because the EMT model predicts little variation in V_S at low hydrate saturation. This complex behavior indicates that the EMT approach may not be suitable for estimating gas hydrates at low hydrate contents. Therefore, we consider the BGTL approach is more appropriate for estimating gas hydrates and free gas in the region where low saturation of gas hydrate is expected.

Acknowledgments

We are thankful to the Director, NGRI for his kind consent to publish this work. The Ministry of Earth Sciences and the Department of Science and Technology, New Delhi are acknowledged for their financial support. Two anonymous reviewers are thanked for their useful comments and suggestions.

References

1. K. A. Kvenvolden, *Geol. Soc. London Spec. Pub.* **137** (1998) 9–30.
2. E. D. Sloan, Marcel Dekker Inc., New York, 1990.
3. A. V. Milkov, *Earth Sci. Rev.* **66** (2004) 183–197 .
4. J. B. Klauda and S. I. Sandler, *Energy and Fuels* **19** (2005) 459–470.
5. Y. F. Makogon, S. A. Holditch and T. Y. Makogon, *J. Petrol. Sci. Eng.* **56** (2007) 14–31.
6. K. Andreassen, P. E. Hart and M. MachKay, *Mar. Geol.* **137** (1997) 25–40.
7. R. Zoeppritz, *Erdbebenwellen VIII B, Goettinger Nachrichten* **1** (1919) 66–84.
8. E. Demirbag, C. Çoruh and J. C. Costain, *Soc. Expl. Geophys.* (1993), pp. 287–302.
9. M. Ojha and K. Sain, *Mar. Petrol. Geol.* **25** (2008) 637–644.
10. M. W. Lee, D. R. Hutchinson, T. S. Collett and W. P. Dillon, *J. Geophys. Res.* **101** (1996) 20347–20358.

11. J. Dvorkin and A. Nur, *Geophys.* **61** (1996) 1363–1370.
12. C. Ecker, J. Dvorkin and A. Nur, *Geophys.* **63** (1998) 1659–1669.
13. M. Helgerud, J. Dvorkin, A. Nur, A. Sakai and T. Collett, *Geophys. Res. Lett.* **26** (1999) 2021–2024.
14. M. W. Lee, *Geophys.* **67** (2002) 1711–1719.
15. S. T. Chand, A. Minshull, D. Gei and J. M. Carcione, *Geophys. J. Int.* **159** (2004) 573–590.
16. K. Sain, T. A. Minshull, S. C. Singh and R. W. Hobbs, *Mar. Geol.* **164** (2000) 37–51.
17. S. R. Fowler, R. S. White and K. E. Louden, *Earth Planet. Sci. Lett.* **75** (1985) 427–438.
18. T. A. Minshull and R. S. White, *J. Geophys. Res.* **94** (1989) 7387–7402.
19. L. F. Athy, *Am. Assoc. Pet. Geol. Bull.* **14** (1930) 1–24.
20. J. Dvorkin, D. Moos, J. L. Packwood and A. Nur, *Geophys.* **64** (1999) 1759–1759.
21. M. Warner, *Tectonophys.* **173** (1990) 15–23.
22. W. J. Ostrander, *Geophys.* **49** (1984) 1637–1649.
23. R. D. Hyndman and G. D. Spence, *J. Geophys. Res.* **97** (1992) 6683–6698.
24. R. E. Sheriff and L. P. Geldert, Cambridge University Press, New York, Vol. 1 (1995).
25. G. Yu, *Geophys.* **50** (1985) 2697–2708.
26. M. Ojha and K. Sain, *Mar. Geophys. Res.* **28** (2007) 101–107.
27. J. Korenega, W. S. Holbrook, S. C. Singh and T. A. Minshull, *J. Geophys. Res.* **102** (1997) 15345–15365.
28. T. A. Minshull, S. C. Singh and G. K. Westbrook, *J. Geophys. Res.* **99** (1994) 4715–4734.
29. S. C. Singh and T. A. Minshull, *J. Geophys. Res.* **99** (1994) 24221–24233.
30. J. P. Castagna and S. W. Smith, *Geophys.* **59** (1994) 1849–1855.
31. J. P. Castagna, H. W. Swan and D. J. Foster, *Geophys.* **63** (1998) 948–956.
32. J. M. Carcione and U. Tinivella, *Geophys.* **65** (2000) 54–67.
33. T. Yuan, G. D. Spence, R. D. Hyndman, T. A. Minshull and S. C. Singh, *J. Geophys. Res.* **104** (1999) 1179–1191.
34. M. P. Chen, M. Riedel, R. D. Hyndman and S. E. Dosso, *Geophys.* **72** (2007) C31–C43.
35. R. T. Shuey, *Geophys.* **50** (1985) 609–614.

Advances in Geosciences
Vol. 18: Ocean Science (2008)
Eds. Jianping Gan et al.
© World Scientific Publishing Company

ESTIMATION OF THE EQUATORIAL PACIFIC SALINITY FIELD USING OCEAN DATA ASSIMILATION SYSTEMS

Y. FUJII*, S. MATSUMOTO and M. KAMACHI

Ocenographic Research Department, Meteorological Research Institute
Nagamine, Tsukuba, Ibaraki 305-0052, Japan
* yfujii@mri-jma.go.jp
www.mri-jma.go.jp

S. ISHIZAKI

Office of Marine Prediction, Japan Meteorological Agency
Ohtemachi, Tokyo 100-8122, Japan
s_ishizaki@met.kishou.go.jp

Efforts at improving salinity fields in the equatorial Pacific in ocean data assimilation systems are reviewed in this paper. Many ocean assimilation systems did not modify model salinity fields directly using observation data until recently. In those systems, salinity is adjusted to the temperature fields to which observation data is assimilated. Recent high level of interest on salinity in the equatorial Pacific and the rapid increase of salinity observations by Argo floats, however, urged researchers to develop multivariate schemes considering relations between temperature and salinity. These schemes can estimate salinity even if without salinity observations. They also avoid destabilization of density stratification caused by unbalanced temperature correction with salinity. The schemes improve reproducibility of the salinity maximum in the tropical subsurface layer, affecting temperature and current fields. Spurious vertical circulation in the equatorial zonal section in many data assimilation systems also degrades salinity fields seriously. This circulation can be removed by correcting bias in the model pressure field.

1. Introduction

The temperature field in the equatorial Pacific has great variability provoked by El Niño — Southern Oscillation (ENSO) (e.g. Ref. 1). This variability affects the worldwide climate in the coupled ocean-atmosphere system (e.g. Ref. 2). Since it has been considered that understanding of the temperature variability is essential for realization of reliable seasonal-interannual climate prediction, the temperature field is intensely observed by TAO/TRITON array[3] and studied by many researchers before (e.g. Refs. 4-6). Theories of the ENSO mechanism are also formulated based on consideration of the temperature variability. For example, the delayed oscillator theory,[7,8] the recharge theory,[9] and the theory of the western Pacific oscillator[10] regard a rise of Sea Surface Temperature (SST) in an El Niño period as a result of thermocline deepening. The advective-reflective oscillator theory[11] considers that the SST variability is caused by displacement of the warm water pool in the western equatorial Pacific.

On the other hand, most researchers have not paid much attention to the salinity field until the end of 1990s. Salinity is, however, a main factor determining seawater density as well as temperature, and therefore acts on variety of oceanic motions including the general ocean circulation and local mixing near the surface. This implies that salinity can affect the temperature field and climate variability in the coupled ocean-atmosphere system.

Why did they ignore that salinity effect? One reason is a lack of knowledge on variability of the salinity field. This lack is caused by the difficulty of salinity observation. Another reason is that they believed the salinity effect on density was secondary. Effects of salinity were considered more carefully in subarctic regions because temperature stratification is weak and salinity has a fundamental role in the vertical density structure. A most part of density stratification in the equatorial ocean is, however, achieved by the temperature distribution. They, therefore, could not imagine that salinity affected the ocean state as well as temperature did in this region.

Recent observations, however, suggest that the near-surface salinity field in the equatorial Pacific has much greater variability than imagined before. There is a sharp salinity front between fresh (low salinity) water (< 34.5 psu) in the western equatorial Pacific and high salinity water (> 35.0) associated with the equatorial upwelling in the central equatorial Pacific. Reference 12 reported that the salinity gradient in the front was more than 1.5 psu over 200 km, and that the front traveled more than 3000 km in a few months. The displacement of the salinity front is focused upon recently. The front is located in the Eastern Warm Pool Convergence Zone (EWPCZ)[13,14] and denotes the eastern edge of the warm water pool in the western equatorial Pacific. The position of EWPCZ is directly connected to the advective-reflective oscillator theory.[11] The salinity front is a good indicator of EWPCZ because it is not easy to detect from the temperature field.

Effects of salinity on temperature and current fields should be also noted around the salinity front. Reference 15 pointed out that the horizontal gradient of the near surface salinity field across the front enhances an eastward current and advection of the warm water. Formation of the barrier layer (the isothermal layer in which salinity stratified)[16] is another important feature associated with the salinity field there. Former studies show[17,18] that the barrier layer is generated by subduction of the eastern high salinity water under the western fresh water caused by South Equatorial Current (SEC). The barrier layer can affect surface currents by thinning the mixed layer and concentrating the effect of wind stress (e.g., Refs. 19, 20). The barrier layer is also considered to induce a rise of temperature in the mixed layer because it prevents warm water at the surface from mixing with cold water in the thermocline. This tendency is confirmed in the equatorial Pacific by former studies.[21-23] Some studies[19,20,24] further confirmed the impact of the barrier layer on onsets of El Niños using Coupled ocean-atmosphere General Circulation Models (CGCMs).

The subsurface salinity maximum associated with South Pacific Tropical Water (SPTW) is another important feature in the salinity field in the equatorial Pacific. SPTW is warm and salty water generated in the subtropical South Pacific, and advected to the subsurface layer in the equatorial Pacific by the subtropical overturning cell. Few studies have examined variability of SPTW before. Reference 25 reported that the salinity maximum of SPTW increases in La Niña periods. Intrusion of SPTW to the northern hemisphere in the early stage of El Niño period is pointed out in Refs. 26–28. It should be noted that SPTW constitutes the base of the barrier layer and also upwells in the central Pacific. The variability of SPTW is therefore possible to affect subsurface temperature and other ocean fields.

It is desirable to compose time series of three-dimensional distribution of the salinity field for understanding of the mechanism on the variability and its impact to the climate system. Ocean data assimilation systems are a powerful tool for that purpose. They synthesize variety of observation data under constraints of ocean general circulation models, and reproduce realistic ocean state and variations, including the salinity field. Actually, salinity was not carefully estimated in data assimilation systems until recently mainly because of the shortage of the observations. Recent high level of interest on salinity in the equatorial Pacific urged researchers to develop more sophisticated schemes for accurate salinity estimation[29] as well as the rapid increase of salinity observations by Argo floats.[30,31] Those systems start to provide time series of realistic salinity fields and new insight on its influence on the ENSO.

Ocean data assimilation systems have another important role: they are used to provide ocean initial condition with CGCMs in ENSO forecasting. Accurate representation of the salinity field in the equatorial Pacific is likely to improve the forecast skill of ENSO. For example, a potential impact of knowledge of the salinity field on the forecast skill is demonstrated statistically in Ref. 32. Potential impacts of the barrier layer on onsets of El Niños studied in Refs. 19, 20, 24 also imply importance of accurate salinity fields for ENSO forecasting. Actually, the impact of improved salinity fields on the forecast skill is demonstrated in Ref. 33. Improving salinity representation is, thus, an essential agenda for the people who deal with ENSO forecasting.

This review aims to summarize recent developments of ocean data assimilation schemes which improve representation of the salinity field in the equatorial Pacific. In Sec. 2, we summarize the multivariate schemes that estimate salinity fields using the Temperature-Salinity (T-S) relation in data assimilation systems. We call this sort of schemes "T-S scheme" in this paper. The impact of T-S schemes on the equatorial Pacific is demonstrated in Sec. 3. In Sec. 4, we describe on a spurious vertical circulation found in the equatorial Pacific in many ocean data assimilation systems. This circulation severely violates equatorial ocean fields, and it is thus essential to mitigate the circulation for producing realistic salinity estimation. This paper is concluded in Sec. 5.

2. Developments of T-S Schemes

Statistical analysis of static ocean fields (e.g., optimal interpolation, and Three-Dimensional Variational (3DVAR) method) is currently adopted in most of operational ocean data assimilation systems for ENSO monitoring and forecasting. The analyzed static fields are used for correcting model fields in some sort of insertion techniques such as direct insertion, nudging, and incremental analysis updates.[34] In those systems, temperature fields are necessarily estimated in the statistical analysis, and used for the direct correction of model temperature fields because temperature has a primal effect on the density field and the relatively large number of temperature observations makes the analysis easier. Salinity fields had, however, not been estimated in the analysis in many ocean data assimilation systems until recently because the number of salinity observation data available for the analysis was very small. The model salinity fields had been, then, just adjusted to the modified temperature fields through the model physics there.

Some studies, had, however, pointed out that appropriate treatments of salinity are essential for controlling model fields realistically because of the following two reasons. One is the effect of salinity variations on Sea Surface Height (SSH) and pressure fields through the contribution to density variations. This issue was firstly addressed by Ref. 35. This paper shows substantial contribution of salinity to the horizontal variation of SSH fields in the tropical oceans, and demonstrated that the salinity contribution affects current fields. Reference 36 then shows from observation data that in 1996 variability of salinity in the western equatorial Pacific can account for 5–10 cm of SSH variability. Reference 37 also shows that, in the case of analyzing temperature alone without considering salinity anomaly, the use of observed SSH in addition to *in situ* temperature data can increase the analysis errors because spurious temperature anomaly is estimated in order to compensate contribution of salinity to SSH.

The other reason is the possibility to destroy density stratification, which is firstly addressed by Ref. 38. Temperature and salinity decrease with depth in many areas of the tropical and subtropical oceans. Upward shift of water mass induces cold and low salinity anomaly in those areas. In the analysis of temperature alone, the cold anomaly is reproduced but the low salinity anomaly is missed, resulting in density larger than real. This inconsistency between temperature and salinity tends to cause density instability and spurious vertical mixing. Actually, this effect severely diminishes the salinity maximums in the subsurface layers in the tropical oceans as demonstrated in the next section.

Estimating salinity together with temperature in the statistical analysis is, thus, essential for improving ocean data assimilation systems. Since salinity observation was very sparse until recently, some sort of schemes estimating salinity from available observation data was required. Many studies therefore had suggested schemes estimating salinity from temperature and SSH observations. For example, salinity fields, as well as temperature fields, are estimated using regression coefficients of their anomaly with SSH anomaly in Refs. 51 and 52. This method can, however, only correct model salinity profiles in the same vertical pattern, and is incapable of estimating salinity anomaly in the place where the anomaly is not reflected in SSH variations. In addition, they do not effectively use information included in temperature observations for estimating the salinity field.

Most of other schemes depend on T-S relations. Ocean water masses usually have specific temperature and salinity. While distribution of ocean water masses is vigorously changed by fluctuation of the current field, responses to the wind stress, and ocean wave activities, they almost conserve temperature and salinity, and thus preserve T-S relation, except in mixed layers. This implies that salinity is determined via T-S relation if we can identify water masses from temperature. Temperature is thus powerful information for estimating the salinity field. As mentioned in the previous section, we call schemes using T-S relations "T-S schemes" in this paper.

Here, we classify T-S schemes into three categories. The schemes in the first category uses climatological T-S relations typically calculated from World Ocean Atlas.[43,44] Synthetic salinity profiles derived from the T-S relation is employed as bogus salinity observations in the statistical analysis in some ocean data assimilation systems.[45-47] The T-S relation is used for constructing a background error covariance matrix in Ref. 48. Salinity fields are estimated through the T-S relation first and then improved by using SSH observations in Refs. 53 and 54. It is also adopted as a constraint in the 3DVAR analysis in Refs. 49 and 50. It should be however noted that the real T-S relations are fluctuated and these schemes cannot reproduce those fluctuations.

The second category includes T-S schemes using vertical shifts of water masses in background (model) T-S profiles. A method estimating temperature and salinity together from SSH anomaly using the vertical shifts is proposed in Ref. 55. Reference 56 also proposed a method estimating the salinity fields from temperature observation using the vertical sifts, which is applied in Ref. 57. These methods are, then, combined in Ref. 33. The methods preserve the T-S relation of water columns in the background state (model fields), and never generate density inversion. It can however not change the T-S relation even if it is not realistic. This scheme is not therefore capable to correct the model T-S relation, which can be a weakness as pointed out in Ref. 46. It is also noted that these schemes cannot correct the model salinity field in the mixed layer. The scheme is improved for assimilating salinity observation additionally in Ref. 58, and adopted in the operational system in European Centre for Medium-Range

Weather Forecasting (ECMWF).[59] The idea of the vertical shifts of T-S profiles also applied to construct a background error covariance matrix in Ref. 41.

The third category includes methods using coupled T-S Empirical Orthogonal Function (EOF) modal decomposition in order to reconstruct vertical profiles of temperature and salinity together. This method is firstly employed for estimating vertical salinity profiles from temperature and SSH observation in the equatorial Pacific[60,61] and the sea east of Japan.[62] It was also used for analyzing three dimensional fields, and successfully reproduces the eastward shift of the barrier layer in the onset of the 1997–98 El Niño in Ref. 22. It then applied to a Mediterranean ocean data assimilation system in Ref. 63 and global and North Pacific ocean data assimilation systems in Ref. 42. T-S EOFs include relations not only between temperature and salinity in the same levels, but also between those in different levels. For example, some EOF modes in the equatorial Pacific represent relation between the thermocline and sea surface salinity, controlling the barrier layer.[22] Salinity in the mixed layer can be estimated through these modes, which is a superior feature to the T-S scheme using vertical shifts of model profiles. The EOF modes are estimated from *in situ* temperature and salinity profiles in those methods. The feasibility is therefore restricted by the availability of observation profiles. On the other hand, EOF modes are calculated from an interannual model simulation in Ref. 64. In addition, an element of SSH variation is included in the EOF modes in order to assimilate SSH observations in the study. The Monte Carlo method is also employed to estimate EOF modes including elements of T, S and current fields in Ref. 40. These methods are not affected by the availability of observations although the EOF modes calculated in this method does not perfectly reflect the real variability.

Recent increase of salinity observation by Argo floats improves salinity fields analyzed by ocean data assimilation systems (e.g., Ref. 47) as well as the development of salinity analysis schemes described above. It is also expected that recent development of sophisticated data assimilation scheme such as four-Dimesional Variational (4DVAR) method (e.g., Refs. 65 and 66) and Kalman Filter (KF) method (e.g., Ref. 67) can also contribute to the improvement of salinity fields because those schemes can reflect T-S relations optimally in their analysis field through model dynamics (in 4DVAR) or time evolution of error covariances (in KF) without any special treatment.

3. Impact of T-S Schemes

In the previous section, we pointed out that the importance of correcting model temperature and salinity fields together, and described on T-S schemes, that is, schemes estimating the salinity field through T-S relations for the use in correction of model salinity fields. In this section, we summarize improvements by correcting model salinity fields using T-S schemes.

Here, we show the results of two assimilation runs using MOVE-G, that is, a global ocean data assimilation system in JMA,[42] in order to demonstrate the impact of T-S schemes. The horizontal grid spacing in MOVE-G is 1°, with meridional equatorial refinement to 0.3° within 5°S–5°N. It applies coupled T-S EOF modes for the estimation of the salinity field. One assimilation run (CTL) is a regular data assimilation run, while the other run (NOS) is a data assimilation run in which the model temperature field alone is corrected using the analysis result. The model salinity field is only adjusted to the modified temperature field according to model equations in NOS. Data assimilated in both runs are *in situ* temperature and salinity profiles and SSH observed by satellite altimetry. Comparison of these assimilation runs shows the impact of T-S schemes. Similar comparison can be seen in Ref. 42 although the assimilation system is updated.

The large impact of a T-S scheme on the subsurface salinity maximum associated with SPTW in the equatorial Pacific is firstly found by Ref. 57. The salinity maximum is severely reduced in their assimilation run without the T-S scheme, while it is adequately reproduced in the run with the T-S scheme. This reduction of the salinity maximum was also found in other conventional assimilation systems without T-S schemes,[39-42] and is now considered as their common defeat. The difference of salinity fields between CTL and NOS (Fig. 1) also demonstrates this defeat. High salinity associated with SPTW is estimated as being inadequately low in NOS.

Reference 57 also found that salinity is contrarily increased in the intermediate layer below the thermocline in the assimilation result without the T-S scheme, which is also confirmed by Ref. 41. This feature is also confirmed in Figs. 1(b) and 1(d). The salinity less than 34.6 psu exists above 400 m depth in the South Pacific in CTL, but it is scarcely found in NOS. Reference 57, then, concluded that these faults are the result of spurious vertical mixing caused by data assimilation without balancing temperature and salinity. Both temperature and salinity decrease with depth under the salinity maximum in this region. The salinity stratification counteracts

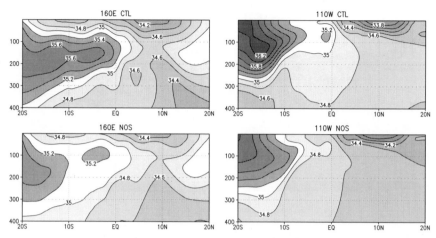

Fig. 1. Meridional vertical sections of climatological salinity fields (psu) at 160°E (left) and 110°W (right) in 1993–2006. Top: CTL, bottom: NOS.

against the density stratification. This means that small reduction of the vertical temperature gradient by data assimilation without adequate salinity modification easily causes unstable density stratification, resulting in spurious vertical mixing between warm and salty water around the salinity maximum and cold and fresh water in the intermediate layer. Thus, the water around the salinity maximum is changed to being fresher, and the water in the deeper layer is changed to being saltier.

This mixing also makes the water around the salinity maximum colder and that in the intermediate layer warmer. The salinity bias caused by unbalanced data assimilation, thus, violates the temperature field. However, this bias is not clear sometimes because constraints to temperature observations in data assimilation systems mitigate it. Reference 57 found substantial warming of the intermediate water, but cooling around the thermocline is not notable in their result. On the contrary, Ref. 42 demonstrated that the cooling around the thermocline is remarkable. It should be noted that the water around the thermocline is advected to the east by the Equatorial Undercurrent (EUC) and upwells in the eastern equatorial Pacific. The near-surface layer in NOS is actually cooler than in CTL because of this bias although SST is rarely affected by the bias because of the assimilation of SST data. The cool bias in the near-surface layer is likely to affect the prediction of SST in the equatorial Pacific.

The temperature field is also violated around the barrier layer, since the absence of the strong salinity maximum prevents substantial formation

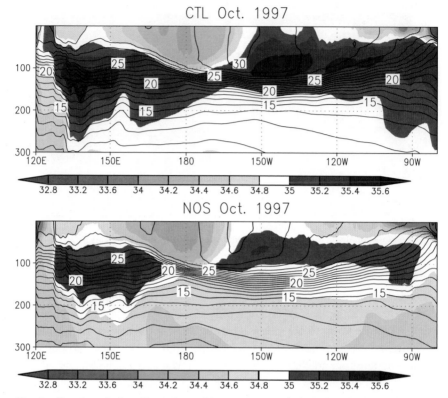

Fig. 2.   Zonal vertical sections of monthly temperature (contour) and salinity (shading, psu) fields along the equator in October 1997. Top: CTL, bottom: NOS. The contour interval is 1°C.

of the barrier layer. Figure 2 shows temperature and salinity fields in the equatorial Pacific in October 1997. In CTL, a substantial salinity maximum is seen at about 100 mm depth around 150°W. The thermocline lies under the maximum and the isothermal layer extends above. This thick isothermal layer is sustained by the barrier layer: the density stratification caused by vertical salinity gradient prevents surface warm water from mixing with cold water in the thermocline. The salinity maximum however weakens and the depth of the salinity maximum is shallower in NOS. This salinity difference caused colder temperature above the thermocline and makes the bottom of the isothermal layer shallower.

Figure 3 shows the variation of the barrier layer thickness and the difference of the warm water heat content between CTL and NOS at the equator. Here, the warm water heat content is defined as the heat content

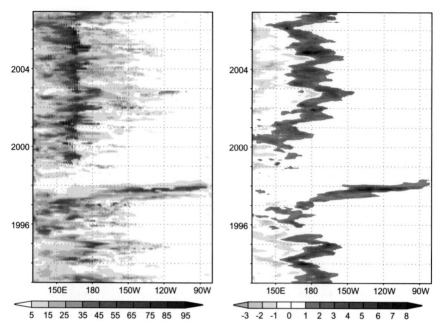

Fig. 3. Longitude-time sections of barrier layer thickness (left, m) and difference of the warm water heat content between CTL and NOS (right, $kcal\,cm^{-2}$).

in water exceeding 28°C. The thick barrier layer is displaced according to the ENSO cycle. It moves to the eastern equatorial Pacific in the large El Nino period, 1997, and temporally disappears after the event. Then, the position of large positive difference of the warm water heat content has a good correspondence with the position of the thick barrier layer. The barrier layer tends to increase the heat content in the warm water by avoiding vertical mixing in CTL. The low salinity bias of SPTW, however, prevents the substantial formation of the barrier layer, which results in the reduction of the warm water heat content in NOS. Salinity correction, thus, improves the subsurface temperature field by estimating the vertical density stratification properly.

Some studies also reported the impact of T-S schemes on velocity fields although the impact depends on ocean data assimilation systems. Reference 39 found an eastward bias in the surface current south of the equator in the eastern equatorial Pacific in their data assimilation system without a T-S scheme. Then, Ref. 41 reported that this bias is removed with a T-S scheme. They concluded that, in the run without the T-S scheme, the higher salinity bias in the intermediate layer south of equator and lower salinity bias in the subsurface around the equator affects the dynamic height errors, resulting in the spurious geostrophic eastward current, and removal of the salinity error by the T-S scheme modifies the current bias.

The influence of T-S schemes on EUC is also reported by some studies, EUC is weakened with the T-S scheme in Ref. 41. Many of ocean data assimilation systems have spurious vertical circulation in the equatorial section as described in the next section. The study shows that the spurious vertical circulation in the equatorial section is reduced with the T-S scheme, and influence of the reduction on heat and salinity budgets in the equatorial section is also addressed. On the contrary, Ref. 42 reported that a T-S scheme intensifies EUC together with its downward shift. They inferred that this difference is caused by the downward shift of the pycnocline and the reduction of the vertical density gradient induced by the spurious vertical mixing.

In summarizing this section, correcting model temperature without balancing salinity causes spurious vertical mixing, violating ocean state in conventional ocean data assimilation systems. This defeat is, then, improved by applying T-S schemes.

4. Spurious Vertical Circulation in the Equatorial Pacific

Another remarkable issue for salinity estimation in the equatorial Pacific is spurious vertical circulation in the equatorial section. This problem is found in many of ocean data assimilation systems.[39,42,68–70] The circulation is caused by modification of the ocean pressure field through temperature (and salinity) assimilation without balancing with wind stress. Atmospheric data often has large bias in the zonal wind stress field near the equator. The zonal pressure gradient is primarily balanced with the inaccurate wind stress when data assimilation is not applied. Although data assimilation reproduces a more accurate pressure gradient field via ocean observations, the field is not balanced with the prescribed zonal wind stress. As a result, clockwise (counterclockwise) circulation basically occurs in the equatorial vertical section seen from the south if the easterly trade wind is too weak (strong) in the atmospheric data, while other errors in ocean models may influence the circulation. The clockwise circulation causes strong downwelling in the eastern equatorial Pacific, inducing warm and salty bias there. On the contrary, the counterclockwise circulation induces strong equatorial upwelling and decreases temperature severely in the eastern equatorial Pacific. Central and western equatorial Pacific is also affected by spurious upwelling or downwelling caused by the circulation. Dealing with this issue is therefore important for improving temperature and salinity fields in the equatorial Pacific.

In order to reduce the spurious circulation, Ref. 69 proposed the zonal current correction using the meridional derivative of the geostrophic balance equation.[71] Although the method improves the zonal current fields, the spurious circulation is not completely canceled. Reference 41 reported that a T-S scheme reduces the spurious circulation although the effect is likely to depend on systems as described in the previous section. Reference 72 adjusted the amplitude of equatorial wind stress previously using historical temperature and salinity profiles. Reference 39 showed that 4DVAR with constraints of ocean model equations can also remove this defeat.

The most elegant method removing the spurious circulation is modifying the pressure gradient field with a bias correction scheme. The method is initially proposed by Ref. 73 and improved by Ref. 74. In their method, the unbalanced part of the pressure gradient with wind stress is considered as a bias of the system. They estimate the bias in temperature and salinity fields in the statistical analysis, convert it to the pressure bias, and remove it from the model pressure field. Athough a bias correction scheme is also applied in Ref. 75, correction increments are added to the model temperature and salinity fields there. On the other hand, adding the increments to the model pressure field effectively cancels the spurious circulation, and improves temperature and salinity fields significantly in the eastern equatorial Pacific in Refs. 73 and 74.

5. Conclusion

The importance of salinity variability in the equatorial Pacific has been well recognized recently. The displacement of the salinity front in the equatorial Pacific is one of the key features for comprehending ENSO mechanism[11]. The barrier layer induces anomalously high SST in the warm pool,[22,23] and can affect onsets of El Niños.[20,24] Estimating the salinity field in high accuracy in the ocean data assimilation systems are desired for the analysis of the salinity role. It also important for improving seasonal forecast skills.

In those backgrounds, the schemes estimating salinity fields from temperature and SSH observations through T-S relations are developed, and increased the accuracy of salinity in ocean data assimilation systems. They avoid the spurious vertical mixing caused by unbalanced temperature correction with salinity and improve ocean state estimation in the equatorial Pacific. Spurious vertical circulation in the equatorial section caused by imbalance between wind stress and pressure gradient is another serious issue for ocean data assimilation systems in the equatorial regions. Some approaches that remove the spurious circulation are studied recently. Ocean state estimation, including the salinity field, is possible to be improved through those efforts. In addition, the increase of salinity observations by Argo floats and development of sophisticated data assimilation schemes such as 4DVAR and KF method are expected to bring further advance to the study of the equatorial Pacific salinity fields.

Acknowledgments

A part of this study was supported by the Grant-in-Aids for Science Research 19540469 from the Ministry of Education, Culture, Sports, Science and Technology, Japan.

References

1. S. G. Philander, *El Niño, La Niña and the Southern Oscillation* (Academic Press, San Diego, 1990).
2. K. E. Trenberth, G. W. Branstator, D. Karoly, A. Kumar, N.-C. Lau and C. Ropelewski, *J. Geophys. Res.* **103** (1998) 14,291.
3. M. J. McPhaden, A. J. Busalacchi, R. Cheney, J. Donguy, K. S. Gage, D. Halpern, M. Ji, P. Julian, G. Meyers, G. T. Mitchum, P. P. Niiler, J. Picaut, R. W. Reynolds, N. Smith and K. Takeuchi, *J. Geophys. Res.* **103** (1998) 14,169.
4. K. Wyrtki, *J. Geophys. Res.* **90** (1985) 7129.
5. C. S. Meinen and M. J. McPhaden, *J. Climate* **13** (2000) 3551.
6. M. J. McPhaden and D. Zhang, *Nature* **415** (2002) 603.
7. M. J. Suarez and P. S. Schopf, *J. Atmos. Sci.* **45** (1988) 3283.
8. D. S. Battisti and A. C. Hirst, *J. Atmos. Sci.* **46** (1989) 1687.
9. F. F. Jin, *J. Atmos. Sci.* **54** (1997) 811.
10. R. H. Weisberg and C. Wang, *Geophys. Res. Lett.* **24** (1997) 779.
11. J. Picaut, F. Masia and Y. du Penhoat, *Science* **277** (1997) 663.
12. C. Hénin, Y. du Penhoat and M. Ioualalen, *J. Geophys. Res.* **103** (1988) 7523.
13. J. Picaut, M. Ioualalen, C. Menkes, T. Delcroix and M. I. McPhaden, *Science* **274** (1996) 1486.
14. J. Picaut, M. Ioualalen, T. Delcroix, F. Masia, R. Murtugudde and J. Vialard, *J. Geophys. Res.* **106** (2001) 2363.
15. D. Roemmich, D. M. Morris, W. R. Young and J.-R. Donguy, *J. Phys. Oceanogr.* **24** (1994) 540.
16. R. Lukas and E. Lindstrom, *J. Geophys. Res.* **96** (1991) 3343.
17. J. Vialard and P. Delecluse, *J. Phys. Oceanogr.* **28** (1998) 1089.

18. M. F. Cronin and M. J. McPhaden, *J. Geophys. Res.* **107** (2002) 8020.
19. J. Vialard, P. Delecluse and C. Menkes, *J. Geophys. Res.* **107** (2002) 8005.
20. C. Maes, J. Picaut and S. Belamari, *Geophys. Res. Lett.* **29** (2002) 2206.
21. K. Ando and M. J. McPhaden, *J. Geophys. Res.* **102** (1997) 23,063.
22. Y. Fujii and M. Kamachi, *J. Geophys. Res.* **108** (2003) 3297.
23. C. Maes, K. Ando, T. Delcroix, W. S. Kessler, M. J. McPhaden and D. Roemmich, *Geophys. Res. Lett.* **33** (2006) L06601.
24. C. Maes, J. Picaut and S. Belamari, *J. Climate* **18** (2005) 105.
25. W. S. Kessler, *J. Phys. Oceanogr.* **29** (1999) 2038.
26. T. Delcroix, G. Eldin, M. H. Radenac, J. Toole and E. Firing, *J. Geophys. Res.* **97** (1992) 5423.
27. G. C. Johnson, M. J. McPhaden, G. D. Rowe and K. E. McTaggart, *J. Geophys. Res.* **105** (2000) 1037.
28. Y. Fujii, Study of estimating temperature and salinity fields using coupled EOF modes, Ph.D. Thesis, Kyoto University, Kyoto (2003).
29. M. Kamachi, Y. Fujii and X. Zhu, *J. Oceanogr.* **58** (2002) 45.
30. J. Boutin and N. Martin, *IEEE Geosci. Remote Sens. Lett.* **3** (2006) 202.
31. J. W. Gould and J. Turton, *Weather* **61** (2006) 17.
32. J. Ballabrera-Poy, R. Murtugudde and A. Busalacchi, *J. Geophys. Res.* **107** (2002) 8007.
33. J. Segschneider, D. L. T. Anderson, J. Vilard, M. Balmaseda and T. N. Stockdale, *J. Climate* **14** (2001) 4292.
34. S. C. Bloom, L. L. Takacs, A. M. Da Silva and D. Ledvina, *Mon. Weather Rev.* **124** (1996) 1256.
35. N. S. Cooper, *J. Phys. Oceanogr.* **18** (1988) 697.
36. C. Maes, *Geophys. Res. Lett.* **25** (1998) 3551.
37. M. Ji, R. W. Reynolds and D. Behringer, *J. Climate* **13** (2000) 216.
38. R. A. Woodgate, *Ocean Modeling* **114** (1997) 4.
39. J. Vialard, A. T. Weaver, D. L. T. Anderson and P. Delecluse, *Mon. Weather Rev.* **131** (2003) 1379.
40. A. Borovikov, M. M. Rienecker, C. L. Keppenne and G. C. Johnson, *Mon. Weather Rev.* **133** (2005) 2310.
41. S. Ricci, T. Weaver, J. Vialard and P. Rogel, *Mon. Weather Rev.* **133** (2005) 317.
42. N. Usui, S. Ishizaki, Y. Fujii, H. Tsujino, T. Yasuda and M. Kamachi, *Adv. Space Res.* **37** (2006) 806.
43. S. Levitus and T. P. Boyer, *World Ocean Atlas 1994*, Vol. 4: Temperature, NOAA Atlas NESDIS 3 (US Dept. of Commerce, Washington DC, 1994).
44. S. Levitus, R. Burgett and T. P. Boyer, *World Ocean Atlas 1994*, Vol. 3: Salinity, NOAA Atlas NESDIS 3 (US Dept. of Commerce, Washington DC, 1994).
45. K. Sakamoto and M. Ishii, *Weather Service Bulletin*, Vol. 70 (Japan Meteorological Agency, Tokyo (2003), p. 131.
46. C. Sun, M. M. Rienecker, A. Rosati, M. Harrison, A. Wittenberg, C. L. Keppenne, J. P. Jacob and R. M. Kovach, *Mon. Weather Rev.* **135** (2007) 2242.
47. B. Huang, Y. Xue and W. Behringer, *J. Geophys. Res.*, Vol. 113, p. C08002.

48. J. A. Carton, G. Chepurin and X. Cao, *J. Phys. Oceanogr.* **30** (2000) 294.
49. G. Han, J. Zhu, G. Zhou, *J. Geophys. Res.* **109** (2004) C03018.
50. C. Yan, J. Zhu, R. Li, G. Zhou, *J. Geophys. Res.* **109** (2004) C08010.
51. T. Ezer and G. L. Mellor, *J. Phys. Oceanogr.* **24** (1994) 832.
52. M. Kamachi, T. Kuragano, N. Yoshioka, J. Zhu and F. Uboldi, *Adv. Atmos. Sci.* **18** (2001) 767.
53. F. C. Vossepoel, R. W. Reynolds and L. Miller, *J. Atmos. Oceanic Technol.* **16** (1999) 1401.
54. F. C. Vossepoel and D. W. Behringer, *J. Phys. Oceanogr.* **30** (2000) 1706.
55. M. Cooper and K. Haines, *J. Geophys. Res.* **101** (1996) 1059.
56. A. Troccoli and K. Haines, *J. Atmos. Oceanic Technol.* Vol. 16 (2011).
57. A. Troccoli, B. A. Balmaseda, J. Segschneider, J. Vialard and D. L. T. Anderson, *Mon. Weather Rev.* **130** (2002) 89.
58. K. Haines, J. D. Blower, J.-P. Drecourt, C. Liu, A. Vidard, I. Astin and X. Zhou, *Mon. Weather Rev.* **134** (2006) 759.
59. M. A. Balmaseda, A. Vidard and D. L. T. Anderson, *Mon. Weather Rev.* **136** (2008) 3018.
60. C. Maes, D. Behringer, R. W. Reynolds and M. Ji, *J. Atmos. Oceanic Technol.* **17** (2000) 512.
61. C. Maes and D. Behringer, *J. Geophys. Res.* **105** (2000) 8537.
62. Y. Fujii and M. Kamachi, *J. Oceanogr.* **59** (2003) 173.
63. E. Demirov, N. Pinardi, C. Fratianni, M. Tonani, L. Giacomelli and P. De Mey, *Annales Geophisicae* **21** (2003) 189.
64. S. Dobricic, N. Pinardi, M. Adani, A. Bonazzi, C. Fratianni and M. Tonani, *Q. J. R. Meteorol. Soc.* **131** (2005) 3627.
65. S. Masuda, T. Awaji, N. Sugiura, Y. Ishikawa, K. Baba, K. Horiuchi and N. Komori, *Geophys. Res. Lett.* **30** (2003) 16,1868.
66. A. Köhl, D. Stammer and B. Cornuelle, *J. Phys. Oceanogr.* **37** (2007) 313.
67. F. Durand, L. Gourdeau, T. Delcroix and J. Verron, *J. Geophys. Res.* **107** (2002) 8004.
68. M. J. Bell, R. M. Forbes and A. Hines, *J. Mar. Syst.* **25** (2000) 1.
69. G. Burgers, M. Balmaseda, F. Vossepoel, G. J. van Oldenborgh and P. J. Leeuwen, *J. Phys. Oceanogr.* **32** (2002) 2509.
70. M. R. Huddleston, M. J. Bell, M. J. Martin and N. K. Nichols, *Q. J. R. Meteorol. Soc.* **130** (2004) 853.
71. H. L. Bryden and E. C. Brady, *J. Phys. Oceanogr.* **15** (1985) 1255.
72. Y. Fujii, T. Nakaegawa, S. Matsumoto, T. Yasuda, G. Yamanaka and M. Kamachi, *J. Climate* (2009).
73. M. J. Bell, M. J. Martin and N. K. Nichols, *Q. J. R. Meteorol. Soc.* **130** (2004) 873.
74. M. A. Balmaseda, D. Dee, A. Vidard and D. L. T. Anderson, *Q. J. R. Meteorol. Soc.* **133** (2007) 167.
75. G. A. Chepurin, J. A. Carton and D. Dee, *Mon. Weather Rev.* **133** (2005) 1328.

Advances in Geosciences
Vol. 18: Ocean Science (2008)
Eds. Jianping Gan *et al.*
© World Scientific Publishing Company

GEOGRAPHICAL INFORMATION SYSTEM APPLIED TO GEOPHYSICAL DATA TO STUDY GAS HYDRATE[*]

MICHELA GIUSTINIANI[†], DANIELA ACCETTELLA[‡],
UMBERTA TINIVELLA, MARIA F. LORETO
and F. ACCAINO

*Istituto Nazionale di Oceanografia e di Geofisica Sperimentale (OGS),
GDL Dept. Borgo Grotta Gigante 42C, Trieste, 34010, Italy*
[†]*migiustiniani@ogs.trieste.it*
[‡]*infloreto@ogs.trieste.it*

We show an application of geographical information system for mapping the regional distribution of gas hydrates reservoir in the South Shetland continental margin. In this area gas hydrates reservoir was detected by using an integrated approach. The available geophysical information are the following: Multibeam data, seismic images, 2D and 3D velocity and porosity models, 2D and 3D gas phase concentrations, pore pressure information, chirp images, gravity core analysis and CTD data. The first step consisted in collecting and homogenizing the data, which has been organized in a specific database, in order to connect all scientific information acquired in the area. This integrated approach has allowed us to obtain regional information, such as 3D distribution of geothermal gradient.

1. Introduction

In this study, we show an application of Geographic information System (GIS) in order to store available data acquired offshore South Shetland Margin. The dataset has been acquired in the frame of two projects regarding the integration of different dataset for the gas hydrate reservoir characterization and supported by the National Program of Antarctic Researcher (PNRA). One of the main goal of these projects is to obtain a reliable model of this area to understand the origin of the hydrates, to predict possible future scenarios related to the climate change, and to investigate the relationship between tectonic and gas hydrates formation.

---

[*]This work is partially supported by Italian National Program of Research in Antarctica (PNRA).

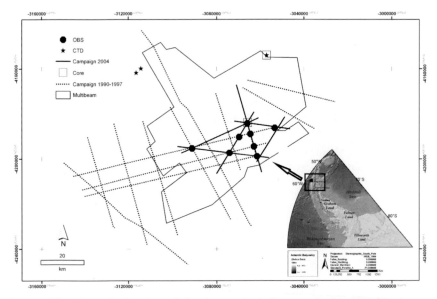

Fig. 1. Map shows the locations of the airgun seismic lines acquired in 2004 (continuous lines, present study), together with those acquired in 1990 and 1997 (dotted lines, cf. Tinivella and Accaino 2000, [4]. CTD measurements (stars), the two coring sites (square), and the border of the Multibeam survey are reported (modified after Tinivella *et al.*, 2007 [5]). The Mercator projection with standard parallel equal to 61°S and spheroid WGS84 was adopted.

In this context, GIS is a very useful tool to develop a database where all information can be stored, managed and integrated. In this paper, we show the preliminary results and the future planned activities.

2. Study Area and Geophysical Dataset

The study area is located in the South Shetland continental margin, as shown in Fig. 1, where the gas hydrates reservoir was detected from seismic data analysis because of the presence of the Bottom Simulating Reflectors (BSRs; i.e. 4). This reflector is generated by the strong contrast of acoustic impedances between sediments bearing gas hydrates and sediments filled with free gas trapped below the hydrate stability zone.

A strong BSR was identified on several multichannel seismic reflection profiles acquired during three cruises performed with the OGS-EXPLORA vessel on the Austral summer 1989/1990, 1996/1997 and 2003/2004 on the South Shetland Margin [4, 5]. During the cruises, more than 1,000 km of multichannel and single channel seismic reflection data were acquired using the following streamers: (1) 3,000 m long 120-channel analogue streamer (during the first two cruises) and (2) 600 m long 48-channel analogue streamer (in the last cruise). The seismic source was an array of air Guns during the first cruise and two GI-guns during the last two ones. In the last two cruises, three-component Ocean Bottom Seismographs (OBSs) were also deployed.

Beyond the seismic data, the following data were acquired:

1. Multibeam acquisition covering an area of about 4,500 km^2. The acquisition was performed by using the Reson Seabat 8,150, which is a full Ocean Depth Multibeam Echo Sounder System, and the processing was performed using the PDS2000 software;
2. Measurements of water velocity profiles by using CTD, two of them in correspondence to sample cores. The adopted instrument was the SBE21, which measures real-time sea temperature and conductivity;
3. sample gravity cores performed in proximity of possible mud volcanoes detected by integrated analysis of Chirp, seismic and Multibeam data;
4. Gravimetric data performed by using marine gravity meter Bodenseewerk. It was always operating and communicating through a serial port to the navigation systems, which recorded and stored the data in digital format. At each port stop, terrestrial measurements were regularly taken by means of a portable Lacoste & Romberg gravity meter so as to link the marine gravity meter measurements to the local reference gravity station and obtain the instrument's drift rate (usually less than 3 mgal/month). The measurements were performed in continuous;
5. CHIRP data acquired by using Sub Bottom Profiler Benthos CAP-6600.

In Fig. 1, the location of the main available data is shown.

3. Seismic Data Analysis

The available seismic data were analysed to extract information about the gas hydrates and their distribution along the margin. We adopted two different approaches to analyse the seismic data.

The multichannel seismic data was analyzed by using iterativelly the pre-stack depth migration with a layer stripping approach to determine the detailed velocity model and the geometry of the main geological structures and in particular the base of the gas hydrate zone [1, 7]. The velocity

model, in fact, gives important information about the spatial distribution
and concentration of the gas hydrate [i.e. 6]. In essential, the adopted
method utilized the output of the pre-stack depth migration (the common
image gathers, CIGs) to determine iteratively an accurate velocity field [3].
If the migrated reflections in CIGs are flat, the correct migration velocity
has been chosen to migrate the data. On the contrary, the not flatness
reflections in the CIGs indicates error in the migration velocity choice.
In this last case, residual moveout analysis measures the deviations by
using the semblance method to correct the curvatures on the CIGs [8].
To obtain an interval velocity for selected layers, we performed the velocity
analysis at selected reflections. At each selected reflection, the update of the
velocity model is performed picking the maximum energy in the semblance
and translating this value in terms of velocity error. Then, we computed a
new migration with the updated velocity model; the procedure stops when
the energy is well focalized around the zero for all selected reflections. In
general, we considered the seafloor, the BSR, a reflector between previous
two and, locally, a reflector below the BSR that, somewhere, represents the
base of the free gas zone (BGR). In Fig. 2, an example of seismic data is
shown; note the presence of the BSR and the BGR.

The obtained velocity models were used to obtain the average velocity
profile, from the seafloor to the BSR, which was used to migrate in depth

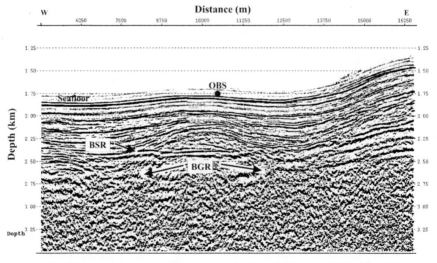

Fig. 2.   Stacked seismic section showing the presence of BSR and BGR (from Accaino
and Tinivella, 2000).

the single channel seismic data. The BSR depth information obtained from the two procedures was interloped to obtained a three dimensional model.

Finally, the BSR depth information was used to extract information about the geothermal gradient. In fact, it is possible to calculate the geothermal gradient (dT/dz) by using the following formula:

$$dT/dz = (T_{BSR} - T_{SEA})/(Z_{BSR} - Z_{SEA}), \qquad (1)$$

where the T is the temperature and the Z is the depth of the BSR and the seafloor. BSR depth was obtained from the seismic data, while seafloor depths derived by the Multibeam data. The seafloor temperature (equal to $0.4°C$) is taken from CTD measurements. The temperature estimate at BSR is based on the dissociation temperature-pressure function of gas hydrates [2]. We considered the following gas composition: methane (90%), ethane (5%) and propane (5%), as indicated by the core analysis [5]. Moreover, we tested three geothermal gradients values: $30°C/km$, $35°C/km$ and $40°C/km$. By using this approach, we evaluated the theoretical depth of the BSR in this area by using the Multibeam data, the water bottom temperature from the CTDs and the geothermal gradient extracted from the seismic data. This information is very useful to estimate regionally the potentiality of the hydrate reservoir.

4. Development of GIS Project

GIS project was developed using ArcGis 9.1 software. The first step consisted in finding available geographic information and geophysical data from bibliography or downloading them from websites. We report the principal source for our scope:

1. Bathymetry of the Antartica obtained from General Bathymetric Chart of the Oceans (GEBCO);

2. Geographic lineaments, such as line coast, downloaded from Antarctic Digital Database (ADD). This information can be downloaded from the website of Scientific Committee on Antarctic Research — SCAR 2008 (http://www.scar.org). ADD is a database containing topographic and geophysical data. All the data has been furnished from many scientific communities involved in Antarctic research.

All information, including data acquired in the two projects above-mentioned, will be projected by using Polar Stereographic Projection with standard parallel equal to $71°S$ and spheroid WGS84, which is the projection adopted by the SCAR. In this way, at the end of the project, we could share our database with all institution involved in Antarctic research.

All available dataset was transformed in a format suitable for GIS. In particular, we transformed the data in the following format: shapefile (internal format of GIS) and grid.

In the following sections, the adopted procedures to integrate the available data are described.

4.1. *Multibeam, CTD and gravimetric data*

Multibeam data has been imported in the GIS project using a file including xyz coordinates projected by using Mercator projection with standard parallel equal to 61° S and spheroid WGS84 and transformed in GRID using some moduli available in "Spatial Analyst" tool. The chosen size of a cell is $200 \times 200\,\mathrm{m}$ (see Fig. 3). This representation of the batimetry is very useful because it has allowed us to recognize important features, such as slumps and mud volcanoes. The area covered by Multibeam acquisition is about $4{,}500\,\mathrm{km}^2$ and the water depth ranges between $-342\,\mathrm{m}$ and $-5{,}500\,\mathrm{m}$.

At this step of the project, the location of CTD and sample cores has been imported in the GIS project by creating a shapefile for each type of data. All available information for each dataset was linked by creating a hyperlink.

The gravimetric data acquired during the cruises allowed us to create a grid that represents the distribution of the Bouger anomalies. The gravimetric data could give an important contribution to improve the structural and geological knowledge in order to understand the gas hydrates formation, even if the preliminary analysis of the data has not given a significant contribution to understand the geological setting.

4.2. *Chirp and seismic data*

Several seismic profiles were acquired and were imported in the GIS project creating a shapefile, which shows the position of each shot.

The detected information, the interpretation of geophysical data, acquisition parameters and the adopted equipment has been inserted in pdf files, creating some cards, which are linked to the shapefiles by using hyperlink.

The integrated analysis of chirp and Multibeam data allowed us to detected 4 unknown mud volcanoes. In order to insert this information

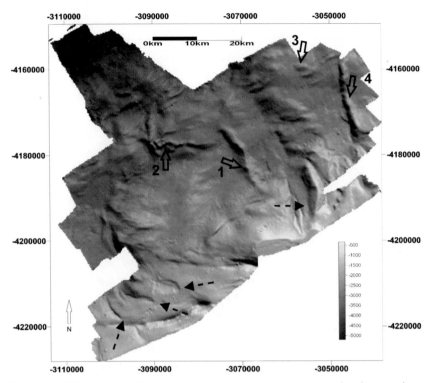

Fig. 3. Multibeam data. The survey allowed us to recognise mud volcanoes (open arrows), collapse troughs (closed arrows) and recent slides (dashed arrows). The numbers indicate the four mud volcano ridges. 1: Vualt and Flop; 2: Grauzaria; 3: Serio; 4: Cjavalz (after Tinivella *et al.* 2007, [5]). The data are projected by using Mercator Projection with standard parallel equal to 61° S and spheroid WGS84.

in the GIS project, a shapefile, which contains all information about the mud volcanoes, has been created. By using the tools available in the GIS, we calculated the dimension of these structures. Moreover, the location of important features, evidenced by the chirp data analyses (i.e. fluid escape, mud volcanoes, and so on), has been stored in a shape file in order to be able to correlate it with other geophysical information. Moreover, we extracted the amplitudes of the seaflooor along all the chirp profile to have information about the variability of the stiffness of the seafloor. Then, the extracted amplitudes were mapped and interpolated to have information about the change of the seafloor reflectivity. This information is very useful to validate our hypothesis about the presence of mud volcanoes and fluid expulsions.

As already explained, we calculated the following information:

1. BSR depth estimated by analyzing the multichannel seismic data;
2. The theoretical BSR depth derived by using the Sloan theory [2].

This information has been transformed in grid adopting the same cell size in order to correlate the different dataset.

4.3. Integration of available dataset

First of all, we compared the Multibeam data with the BSR depth obtained from seismic data in order to understand if the presence of gas hydrates are related to some anomalies, such as geothermal gradient anomalies. In order to reach this goal, we performed some numerical computations between the grids. Being the BSR depth available just along the seismic lines, this information was interpolated. We obtained the theoretical BSR depth by using three different geothermal gradient values. The obtained data was transformed in grid format with a cell size equal to 200 m. Each grid was subtracted to the grid representing BSR depth obtained from seismic data producing three new grids. Assuming an error of the BSR depth calculated from seismic data equal to about 5% and the average BSR depth equal to about 3000 m, we have considered just the values of the differences ranging between about −150 m and 150 m. The analysis of these grids suggested that the regional geothermal gradient could be in a range of 35 and 40° C/km (Fig. 4). The comparison of bathymetry obtained from Multibeam data and the BSR depth obtained from seismic data has given information about anomalies, such as geothermal anomalies.

Other important information, which can be extracted from the GIS, is the correlation between BSR depth anomaly (by using seismic data) and shallow structures (by using chirp and multibeam data). In fact, the BSR depth anomaly is the deviation of the seismic BSR depth with respect to the theoretical BSR depth calculated assuming the regional gradient. To associate the BSR depth anomaly to a geothermal gradient anomaly, it is important to have other geophysical data that can validate this anomaly. In fact, the BSR depth anomaly could be also associated to seismic velocity error, which is difficult to exactly quantify (about 5%). Thus, if there is a correlation between the BSR depth anomaly (i.e. geothermal gradient) and shallow structures (such as fluid escape, mud volcanoes), we can validate the result. In our case, we recognize the geothermal gradient anomaly in proximity of the Vualt mud volcano (see Fig. 2).

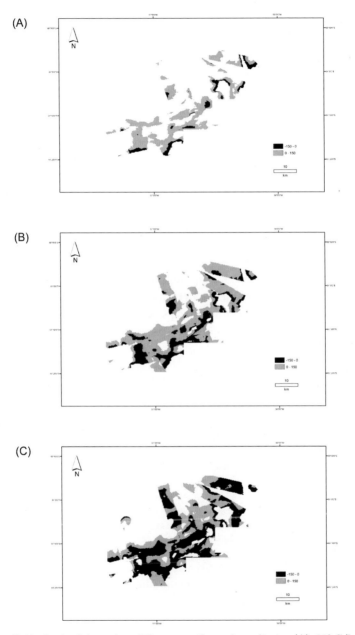

Fig. 4.   Grid obtained by using different geothermal gradients: (A) 30° C/km, (B) 35° C/km and (C) 40° C/km. See text for details. The data are projected by using Mercator projection with standard parallel equal to 61° S and spheroid WGS84.

5. Conclusion

We have shown preliminary results of an application of Geographic Information System to characterize a gas hydrate reservoir. GIS has allowed us to integrate several geophysical data, acquired in the South Shetland Margin (Antarctic Peninsula) where a gas hydrate reservoir is present.

First of all, the available information was imported in the GIS project and georeferenced by using a specific projection. After finding other existing geographic information from various sources, we interpolated 2D information in order to obtain 3D distribution of the same data. For example, BSR depth obtained from seismic data was available just along the seismic lines. By using GIS tools, we interpolated this information obtaining a grid.

The integration of different data has permitted to extrapolate regional information. For example, the analysis of information extrapolated from theoretical model and from seismic data about the BSR depth has allowed us to estimate the regional geothermal gradient that about 35–40° C/km. This information is very useful to correlate geothermal anomalies (identified by deviation of the BSR depth) to other features such as fluid escape.

In the future, the 3D seismic velocity field will be imported in the GIS project in order to increase the knowledge of the distribution and origin of the gas hydrate reservoir present offshore the South Shetland margin.

References

1. M. F. Loreto, U. Tinivella, F. Accaino and M. Giustiniani, *Advanced Geoscience, Proc. AOGS* (2008).
2. J. R. Sloan, Marcel Dekker, Inc., New York, 1998, p. 641.
3. O. Yilmaz, *Society of Exploration Geophysicists, Tulsa Oklahoma* (2001), p. 2027.
4. U. Tinivella and F. Accaino, *Marine Geology* **164** (2000) 13–27.
5. U. Tinivella, F. Accaino and B. della Vedova, *Geomarine Lett.* (2007), doi: 10.1007/s00367-007-0093-z.
6. U. Tinivella, *J. Seis. Explo.* **11** (2002) 283–305.
7. U. Tinivella, M. F. Loreto and F. Accaino, *Geol. Soc. London* (in press).
8. Z. Liu, Ph.D. thesis, Colorado School of Mines, Vol. CWP 168 (1995).

Advances in Geosciences
Vol. 18: Ocean Science (2008)
Eds. Jianping Gan et al.
© World Scientific Publishing Company

CORRELATION BETWEEN GEOLOGICAL STUCTURES AND GAS HYDRATE AMOUNT OFFSHORE THE SOUTH SHETLAND ISLAND — PRELIMINARY RESULTS*

M. F. LORETO

*Istituto Nazionale di Oceanografia e Geofisica Sperimentale,
Borgo Grotta Gigante, 42/C, Sgonico (Trieste), 34135, Italy*
mfloreto@ogs.trieste.it

U. TINIVELLA, F. ACCAINO and M. GIUSTINIANI

*Istituto Nazionale di Oceanografia e Geofisica Sperimentale,
Borgo Grotta Gigante, 42/C, Sgonico (Trieste), 34135, Italy*

Sediments of the accretionary prism, present along the continental margin of the Peninsula Antarctica SW of Elephant Island, are filled by gas hydrates as evidenced by a strong BSR. A multidisciplinary geophysical dataset, represented by seismic data, multibeam, chirp profiles, CTD and core samples, was acquired during three oceanographic cruises. The estimation of gas hydrate and free gas concentrations is based on the P-wave velocity analysis. In order to extract a detailed and reliable velocity field, we have developed and optimized a procedure that includes the pre-stack depth migration to determine, iteratively and with a layer stripping approach method, the velocity field and the depth-migrated seismic section. The final velocity field is then translated in terms of gas hydrate and free gas amounts by using theoretical approaches. Several seismic sections have been processed in the investigated area. The final 2D velocity sections have been translated in gas-phase concentration sections, considering the gas distribution within sediments both uniformly and patchly distributed. The free gas layer is locally present and, consequently, the base of the free gas reflector was identified only in some lines or part of them. The hydrate layer shows important lateral variations of hydrate concentration in correspondence of geological features, such as faults and folds. The intense fluid migration along faults and different fluid accumulation in correspondence of geological structures can control the gas hydrate concentration and modify the geothermal field in the surrounding area.

*This work is supported by Programma Nazionale Ricerche in Antartide (PNRA).

1. Introduction and Tectonic Setting

Gas hydrates is a solid phase composed by water and gas molecules, known as clathrate hydrate [1], which forms under adequate pressure, temperature and gas concentration conditions [2, 3]. The methane is the most common gas trapped within cages of icy molecular structures, also due to its small dimension that strongly stabilizes the crystalline structure of clathrate hydrates. Seismically, the gas hydrate presence is evidenced by a strong reflector called Bottom Simulating Reflector (BSR), which (i) mimics the seafloor reflector [4], (ii) is characterized by the inverse polarity compared to the seafloor, and (iii) cuts the sediment stratifications. The BSR represents the base of the gas hydrate stability zone produced by the strong contrast of acoustic impedance due to the high seismic velocity of the gas hydrate above [5], and to the low velocity of the free gas accumulated below [6]. Higher is the velocity contrast between the two layers and higher will be the amplitude of the BSR [7, 8].

In sub-marine area, stability or instability of gas hydrates within sediments are strongly correlated to geological processes. In fact, some authors have observed that gas accumulation are more diffuse in area with high sedimentation rate and around diapiric structures [9, 10]. Whereas, area showing intense fluid flow are affected by intense tectonics, involving the seafloor because of the presence of mud volcanoes and high geothermal field. Sediments with this characteristic show a spatial discontinuous gas hydrate layer [10, 11]. Hydrate layer variability can result by an interplay of tectonics and sedimentations that favor the formation of gas escape pathways [9–12].

Along the South Shetland margin, see left of Fig. 1, a trench-accretionary prism system is present [13, 14]. This system is the result of the active spreading along the Antarctic-Phoenix ridge stopped about 3.5 Ma ago [15, 16], but subduction continued as a consequence of sinking and roll-back of the subducted slab [14, 17]. Along the continental margin, a narrow accretionary prism has developed, ranging from 20 to 40 km in width. By seismic data interpretation [18] some deformational features were recognized within prism sediments. In particular, the frontal part of the prism is affected by reverse and thrust faults; while, extensional faults further from the trench are detected. Moreover, an important strike slip fault has been interpreted as being related to the Shackleton Fracture Zone, and affecting the internal prism sediments with and orthogonal orientation to the continental margin. Small mid-slope basins are common within the prism, often bounded by extensional faults that locally reach the seafloor

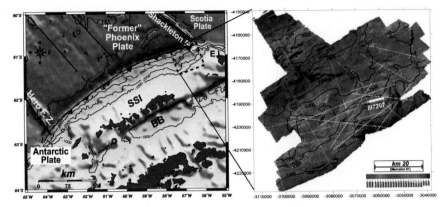

Fig. 1. On the left, schematic shaded-relief morpho-structural map showing mean features [from 19], SSI = South Shetland Is., BB = Bransfield Basin, E.I. = Elephant Is. The arrow indicates the direction of plate convergence. The dashed rectangle indicates the high resolution morpho-bathymetry map of study area, on the right, with the entire available seismic dataset (dotted white lines). Part of line 197207 is here analyzed and indicated with thick white line.

[18]. A clear BSR was interpreted that results in very continuous and with high amplitude in a restricted area located inside of prism sediments (yellow area in Fig. 1, right).

The focus of this work is to estimate the gas hydrate and free gas amounts, trapped within the sediment pore space of the prism presents along the South Shetland margin, and to analyze the variation of the gas concentration related to geological features.

2. Geophysical and Data Processing

2.1. *Seismic dataset*

Seismic data were acquired during the Austral summer 1996/1997 in the frame of a geophysical cruise supported by the Programma Nazionale Ricerche in Antartide (PNRA). The source was composed by two generator-injector (GI guns) with a total volume of 4 liters firing every 25 m. The streamer was 3,000 m long with a hydrophone group interval of 25 m. The sampling interval was 1 ms. To better characterize the area where the BSR is very strong and continuous, a new cruise was carried out to acquire detailed bathymetric data (12 kHz is the acoustic frequency used), sub bottom profile data (7 kHz) and seismic data with a short hydrophone streamer (600 m) during the Austral summer 2003/2004.

The pre-stack depth migration (PreSDM) was applied on selected seismic lines, with long streamer, and a regional velocity fields were obtained with a layer stripping approach and Common Image Gathers (CICs) analysis. A pass-band filter 24–80 Hz was applied to remove frequencies random noise.

2.2. *Applied procedure*

The estimation of gas hydrate amount is derived by using seismic velocities and Biot-Geerstma-Smit theory. The velocity field is obtained with a layer stripping approach [20] and residual moveout analysis on CIGs. The depth velocity model is used to implement the pre-stack depth migration in order to obtain the geometry of the BSR and depth images.

The PreSDM is based on the Kirchhoff algorithm and implemented with the free software Seismic Unix [21]. The method that we used to estimate the velocity by the PreSDM (see details in Liu, 1995), is based on the analysis of the output of the PreSDM. The elaboration of the output allows to determine the velocity error and, then, to update the velocity model [22]. The residual moveout analyses the deviations by using the semblance method in order to correct the curvatures on the CIGs [22]. The reflections in the CIGs are imaged as areas of maximum energy in the semblance. Therefore, a nonzero r-parameter means the presence of residual moveout and consequently an incorrect migration velocity. The update of the velocity model is performed picking the r-parameter in correspondence of the maximum energy in the semblance, and translating the value of r-parameter in terms of velocity error. Then, we computed a new migration with the updated velocity model; the procedure stops when the energy is well focalized around the r-parameter equal to zero for all selected reflections.

Following this method, we implement the first iteration of the PreSDM with a constant velocity model, constructed with the water velocity of 1,465 m/s. After few iterations we fixed the water velocity and started with the velocity inversion of the second layer. The hydrate layer corresponds, in our case, to the third inverted layer for which we obtained the correct velocity values after many iterations. A final velocity depth model, see Fig. 2, is composed by 4 layers: the water layer, the gas hydrate layer, a layer between these two (called first layer), and a deep layer in which we inserted a velocity gradient equal to 0.6 1/s. Then, the model was smoothed to attenuate lateral and vertical velocity variations.

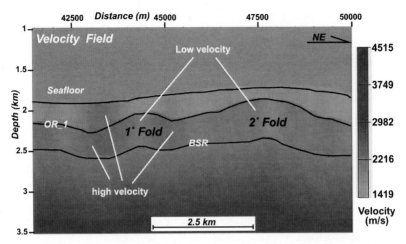

Fig. 2.   Final depth velocity model of part of the line 197-207 (see location in Fig. 1) derived by CIGs analysis and PreSDM method. Horizon 1 is indicated with OR_1.

The gas phase concentration of the line 197-207 was obtained applying the procedure described in Tinivella [23, 24], which translates the velocity anomalies in terms of gas phase concentration. More detail about the procedure applied are reported in Tinivella *et al.* [25]. We produced a gas hydrate concentration section of the entire line, considering a uniform distribution of the free gas within the pore space. Here, we show only the area where folds are present, see Fig. 3.

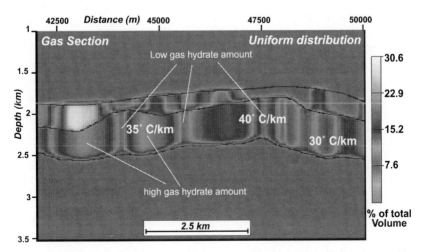

Fig. 3.   Gas hydrate section of part of the line 197-207, derived by Sloan theory. Along the hydrate layer, the estimated geothermal gradient is indicated.

3. Analysis of Results

3.1. *The pre-stack depth migration section*

The final PreSDM section was performed by using the final velocity model. In Fig. 4, a part of line 197-207 is shown. It is a really interesting area where sediments are deformed by folds and faults and a strong BSR is present. From 42.5 to 45.1 km and from 45.2 to 49 km distances, two clear syncline-anticline structures are located. Between the two folds, at a depth of few hundred meters below seafloor, an incipient ramp fault is interpreted, see Fig. 4. Deformation of sediments seem confined below the seafloor, as suggested by the real continuity of the seafloor reflector.

Sediments, along the hinge of the folds, show lateral variations of the seismic characters, such as frequency and amplitude. These variations could be associated to fractures that usually are present within folded sediments, and/or to seismic migration effects.

On the back of the folds are present little sedimentary basins corresponding to the synclinal segment of the structures. Laterally, the basin sediments are truncated in correspondence of the seafloor, probably due to a likely combination of up-welling tectonics and deep water currents.

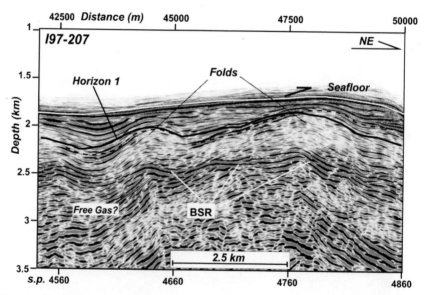

Fig. 4. PreSDM section of part of line 197-207, Fig. 2. The dotted black line indicates the interpretation of folds with NE vergence, as indicated by black arrow. Vertical exaggeration 2:1.

Along these segments the BSR results sub-parallel to the seafloor showing an average depth of about 580 m below the seafloor.

3.2. *Lateral velocity and gas concentration variations*

The gas hydrate layer shows lateral velocity variations ranging between 1,800–2,360 m/s. Two areas with relative low velocities (of about 1930 and 1,800 m/s) are located in correspondence of the hinge of the first and second folds respectively. Laterally, the velocity increases until to reach the maximum value of 2,360 m/s in correspondence of the synclinal segment of the first fold, see Fig. 2. Between the two folds the velocity is not uniform and ranges between 1,960 to 2,100 m/s. The velocity of the first layer shows similar trend, reaching a very high value (2,100 m/s) in correspondence of the syncline. This anomalous high velocity could be linked to the stratigraphic setting. In fact, the syncline geometry of sediments can favor lateral fluid migration, affecting the velocity field, as observed by other authors [27]. Moreover, note that the thickness of the layer is higher in this syncline with respect to the surrounding area; consequently, the interval average velocity between the seafloor and the horizon 1 is higher above the syncline and lower above the anticline, also due to the thickness layer variation. Besides, the velocity above the syncline is higher than expected.

The gas hydrate concentration, see Fig. 3, shows a similar sinusoidal trend of the velocity model. In fact, a very low gas hydrate amount is detected in correspondence of the fold hinges. Note that the gas hydrate amount is estimated in only about 2–3% of total volume on the hinge of the second fold. On the other hand, it is important to underline that the low velocity in this area can also be due to the fractures affecting the fold hinges. Along the sides of the folds, the gas hydrate concentrations increase reaching the maximum value of about 22% of the total volume above the BSR and of about 30% in the fist layer. These high anomalous values are associated not only to the gas hydrate presence, but in most part to the nature of the sediments, which can present low porosity due to syncline geometry effects.

3.3. *Geothermal gradient versus geological features*

To better analyze fluid circulation within sediments the geothermal gradient was estimated. This information was derived comparing the seismic depth of the BSR with the theoretical depth of the BSR, obtained using the Sloan theory [21]. The data used to calculate the theoretical BSR depth includes the seafloor depth, the gas mixture composition (90% of methane, 5% of ethane and 5% of propane), and the sea bottom temperature (equal to 0.4°C [26]). In fact, by comparing the theoretical curves of the BSR with the depth of the BSR, we can estimate the geothermal gradient and its lateral variations. In particular, the geothermal gradient decreases to 30°C/km on the front of the second fold, see Fig. 3, where the BSR is characterized by low amplitude and, locally, disappears. In essential, in this area we obtain an average value of 35°C/km for the geothermal gradient.

4. Conclusions

An integrated analysis of the velocity model, gas amount sections and geothermal gradient can be very useful to determine the relationship between geological features (such as fluid escape, fault, fracture and fold) and gas hydrate presence. In our area, we have detected important velocity and gas hydrate variations along the geological structures, confirming that this approach is very useful to understand the origin and the formation of the hydrate in reservoir area.

Acknowledgments

This work is partially supported by PNRA.

References

1. S. L. Miller, *Proc. Nat. Acad. Sci. USA* **47** (1961) 1798–1808.
2. M. K. MacLeod, *Am. Ass. Pet. Geol. Bull.* **109** (1982) 477–491.
3. T. Minshull and R. White, *JGR* **94** (1989) 7387–7402.
4. T. H. Shipley, M. Houston, R. Buffer, *et al. AAPG Bull.* **63** (1979) 2204–2213.
5. R. Stoll, J. Ewing and G. Bryan, *JGR* **76** (1971) 2090–2094.
6. W. F. Murphy, *JGR* **89** (1984) 1549–1559.
7. J. J. Ewing and C. H. Hollister, *DSDP* **XI** (1972) 951–973.
8. W. S. Holbrook, H. Hoskins, W. T. Wood, R. A. Stephen and D. Lizarralde, *Science* **273** (1996) 1480–1483.
9. W. P. Dillon and M. D. Max, *Natural Gas Hydrate in Oceanic and Permafrost Environment* (Kluwer Accademic Publisher, 2000), pp. 61–76.
10. T. Lüdmann and H. K. Wong, *Marine Geology* **201** (2003) 269–286.
11. C. K. Paull, W. Ussler, W. S. Borowski and F. N. Spiess, *Geology* **23** (1995) 89–92.

12. W. S. Holbrook, D. Lizaralde, I. A. Pecher, A. R. Gorman, K. L. Hackwith, M. Hornback and D. Saffer, *Geology* **30** (2002) 467–470.
13. A. Maldonado, R. D. Larter and F. Aldaya, *Tectonics* **13** (1994) 1345–1370.
14. Y. Kim, H.-S. Kim, R. D. Larter, A. Camerlenghi, L. A. P. Gambôa and S. Rudowski, *Geology and Seismic Stratigraphy of the Atlantic Margin* **68** (1995) 157–166.
15. R. D. Larter and P. F. Barker, *JGR* **96** (1991) 19583–19607.
16. R. Livermore, J. C. Balanyá, A. Maldonado, J. M. Martínez, J. Rodríguez-Fernández, C. Sanz De Galdeano, J. Galindo Zaldívar, A. Jabaloy, A. Barnolas, L. Somoza, J. Hernández-Molina, E. Suriñach and C. Viseras, *Geology* **28** (2000) 607–610.
17. M. F. Loreto, B. Della Vedova, F. Accaino, U. Tinivella and D. Accettella, *Ofioliti* **31** (2006) 135–143.
18. E. Lodolo, A. Camerlenghi, G. Madrussani, U. Tinivella and G. Rossi, *Geophys. J. Int.* **148** (2002) 103–119.
19. D. T. Sandwell and W. H. F. Smith, *JGR* **102** (1997) 10039–10054.
20. F. Accaino, G. Bohm and U. Tinivella, *Firts Break* **23** (2005) 39–44.
21. E. D. Jr. Sloan (New York, 1998), p. 705.
22. Z. Liu, Ph.D. thesis, Colorado School of Mines, CWP (1995), p. 168.
23. U. Tinivella, *Bollettino di Geofisica Teorica ed Applicata*, Vol. 40 (1999).
24. U. Tinivella, F. Accaino and A. Camerlenghi, *Marine Geophys. Res.* **23** (2002) 109–123.
25. U. Tinivella, M. F. Loreto and F. Accaino, *Geol. Soc. London* (in press).
26. U. Tinivella, F. Accaino and B. Della Vedova, *Geomarine Lett.* Vol. 28, pp. 97–106.
27. C.-C. Lin, A. T.-S. Lin, C.-S. Liu, G.-Yu Chen, W.-Z. Liao, and P. Schnurle, Geological controls on BSR occurrences in the incipient arc-continent collision zone off southwest Taiwan, *Marine and Petroleum Geology* (2008), doi:10.1016/j.marpetgeo.2008.11.002.